Finance, Innovation and Geography

The overarching aim of *Finance, Innovation and Geography: Harnessing Knowledge Dynamics in German Biotechnology* is to explore linkages between geographies of finance and relational geographies of innovation. This is achieved by questioning how investment activities affect the unfolding of innovations and in turn are affected by it.

This book focuses on biotechnology innovation processes from the perspective of relational economic geography. It reconstructs the unfolding in time and space of eight innovations in German biotechnology. Each one is represented in a qualitative case study. The analysis focuses on the relational work of building, transforming, ending and replacing of collaborative relationships and organizational arrangements surrounding emergent innovations – including investment relations and relational work by investors. In this way, the contribution of investors to unfolding innovations is studied with sensitivity to context and situated interactions. The geography of these dynamics is conceptualized by drawing on the recent literature on relational proximity and distance as well as ideas of materiality and space.

This book provides a unique perspective, and shows that innovation paths are strongly interwoven with local and temporary opportunities as well as crises, and that investment is embedded in these dynamics. This is essential reading for students and academics of both economics and innovation.

Felix C. Müller is an economic geographer affiliated with the Leibniz Institute for Research on Society and Space (IRS) in Erkner near Berlin. He has researched dynamics of knowledge and value creation from the perspectives of relational economic geography as well as cultural economy.

Routledge Studies in Innovation, Organizations and Technology

Strategic Marketing for High Technology Products
An Integrated Approach
Thomas Fotiadis

Responsible Research and Innovation
From Concepts to Practices
Edited by Robert Gianni, John Pearson and Bernard Reber

Technology Offsets in International Defence Procurement
Kogila Balakrishnan

Social Entrepreneurship and Social Innovation
Ecosystems for Inclusion in Europe
Edited by Mario Biggeri, Enrico Testi, Marco Bellucci, Roel During,
Thomas Persson

Innovation in Brazil
Advancing Development in the 21st Century
Edited by Elisabeth Reynolds, Ben Ross Schneider and Ezequiel Zylberberg

Strategic Renewal
Core Concepts, Antecedents, and Micro Foundations
Edited by Aybars Tuncdogan, Adam Lindgreen, Henk Volberda, and
Frans van den Bosch

Service Innovation
Esam Mustafa

Innovation Finance and Technology Transfer
Funding Proof of Concept
Andrea Alunni

Finance, Innovation and Geography
Harnessing Knowledge Dynamics in German Biotechnology
Felix C. Müller

For more information about the series, please visit www.routledge.com/Routledge-Studies-in-Innovation-Organizations-and-Technology/book-series/RIOT

Finance, Innovation and Geography

Harnessing Knowledge Dynamics in German Biotechnology

Felix C. Müller

Routledge
Taylor & Francis Group

LONDON AND NEW YORK

First published 2019
by Routledge
2 Park Square, Milton Park, Abingdon, Oxon OX14 4RN

and by Routledge
605 Third Avenue, New York, NY 10017

First issued in paperback 2020

Routledge is an imprint of the Taylor & Francis Group, an informa business

© 2019 Felix C. Müller

British Library Cataloguing in Publication Data
A catalogue record for this book is available from the British Library

Library of Congress Cataloging-in-Publication Data
A catalog record has been requested for this book

ISBN 13: 978-0-367-73018-5 (pbk)
ISBN 13: 978-0-8153-9549-2 (hbk)

Typeset in Times New Roman
by Taylor & Francis Books

Contents

Illustrations

Figures

Tables

Acknowledgements

This book is based on my PhD thesis titled 'How Money Tames Innovation. A Dynamic Relational Geography of Investment in German Biotechnology', which I submitted and defended at Freie Universität Berlin. I would like to thank my supervisor and department leader Oliver Ibert for his mentorship, constructive feedback and many highly fruitful discussions. I further wish to thank my external examiner Jane Pollard of Newcastle University for her critical and thorough reading of my work as well as her valuable advice. Both researching investment relations in biotech innovation processes and writing this book were made possible by my employment at the Leibniz Institute for Research on Society and Space (IRS) in Erkner. Both key conceptual developments and important parts of the fieldwork were embedded in two consecutive, institutionally funded 'lead projects' at the research department 'Dynamics of Economic Spaces' at IRS (2009–2014). I benefited greatly from the productive and inspiring atmosphere at my department, and I would like to thank the Institute as well as my colleagues for their continued support. In particular, I wish to thank Verena Brinks and Suntje Schmidt for the highly productive collaboration on several projects, as well as countless stimulating and encouraging conversations. I also want to thank the reviewers of my manuscript for their critical reading and valuable comments. Most of all I want to thank my partner Jacqueline who accompanied me through the years I devoted to this work and shared many of the associated burdens with me. I am truly grateful for everything I received in these past years.

Abbreviations

ANT	Actor-Network Theory
BMBF	Bundesministerium für Bildung und Forschung (Federal Ministry for Science and Education)
BMWi	Bundesministerium für Wirtschaft und Energie (Federal Ministry of Economic Affairs and Energy)
CEO	chief executive officer
CFO	chief financial officer
CMO	contract manufacturing organization
CRO	clinical research organization
CSO	chief scientific officer
CVC	corporate venture capital
DAI	Deutsches Aktieninstitut (German Stock Institute)
DtA	Deutsche Ausgleichsbank (German Equalisation Bank)
EMA	European Medicines Agency
ERP	European Recovery Program ('Marshall Plan')
FDA	Food and Drugs Administration
FIPCo	fully integrated pharmaceutical company
GLP	good laboratory practices
GMO	genetically modified organism
GMP	good manufacturing practices
HGP	Human Genome Project
HTGF	High-Tech Gründerfonds (High Tech Entrepreneurs Fund)
IBB	Investitionsbank Berlin (Berlin Development Bank)
IP	intellectual property
IPO	initial public offering
IPR	intellectual property right
KfW	Kreditanstalt für Wiederaufbau (Reconstruction Loan Corporation)
KWG	Kreditwesengesetz (Credit System Act)
M&A	mergers and acquisitions
MPG	Max Planck-Gesellschaft (Max Planck Society)
MPI	Max Planck-Institut (Max Planck Institute)
NIH	National Institutes of Health

NRW	North-Rhine-Westphalia
PCR	polymerase chain reaction
R&D	research and development
RVCF	regional venture capital fund
STS	Science and Technology Studies
TBG	Technologie-Beteiligungs-Gesellschaft mbH (Technology Equity Company)
VC	venture capital

1 Biotech innovation processes, geography and investment

1.1 Two innovations from Berlin

It's the 1990s, over 20 years before a technology called 'Crispr', and in a wave of general technology enthusiasm hopes are high for a biotechnological revolution to arrive. Molecular biology – along with computing and information science – is seemingly about to replace physics and chemistry as the big drivers of innovation. The first dedicated biotechnology companies like Amgen and Genentech have emerged and grown in the United States since the 1980s. 'Second wave' countries, like the UK and later Germany, are adopting policies to foster the growth of biotechnology clusters. The Human Genome Project is conceived and executed, delivering to stunned audiences a spectacular race between a global community of scientists and a tech-savvy entrepreneur backed-up by an army of sequencing robots. Gene therapies appear to be within reach, ending diseases like diabetes and cancer. World hunger is about to be eradicated, too. Francis Fukuyama's idea of 'the end of history' is fresh and the sensation of permanent crisis characterizing the near two decades after 9/11 is far away. Money, computers, brains and most importantly genes are about to turn utopias into reality.

From today's vantage point, the Dotcom-bubble of 2000/2001 and the crisis caused by its bursting is just one of several speculation bubbles. A much bigger one burst in 2008, and fears of other areas of irrational exuberance (Shiller 2015) in crypto currencies, platforms, artificial intelligence or real estate (or, indeed, biotechnology) are palpable. In addition to the dry spell on technology financing following the crisis, biotechnological solutions themselves turned out to be less capable, the challenges far bigger than expected. One biotechnology company CEO interviewed for this book sternly commented:

> [...] the first attempts in gene therapy [...] ran high at the end of the 90s, which were strongly funded by venture capital [...] generally, and which were abruptly interrupted by two things: on the one hand the capital market crisis of 2001 and on the other hand the death of this poor boy in California. [...] He had a genetic disease, and he was plunged into a much too early, completely flatfooted attempt at gene therapy, from which

he died. And this lastingly discredited the whole area of gene therapy as a model for personalized medicine, which as such is good and important. It took exactly ten years until someone dared to attempt a clinical trial with this again. This is the terrible thing about this industry sometimes, that it is simply too much driven by the capital market and also simply too much by news and hype, so that in the end all measures of precaution and all ethical reservations are jettisoned. (Interview 6–1)

It's the late 1990s, and two biotechnology companies are founded in Berlin. Both originate in – separate – Max Planck Institutes. Max Planck Society is Germany's leading non-university research organization, driving cutting-edge basic research, creating new fields of scientific inquiry and producing most of Germany's Nobel Prize winners. Both ventures have one other thing in common: They are both based on innovative ideas which evade the hegemonic gene-determinism of their time and instead turn to the complexities of bio-molecular interactions between DNA strands and their cellular environments.

> At that time, when everybody around us still reasoned that genes alone are in principal the blue-prints of plants and humans, and hoped, once you've only read the genes, you can in principal derive the functions of the genes. And the founders, to whom I belong, as early as twelve years ago, said this would not be enough. Beyond that, you will have to investigate the Gene Functions experimentally. You have to turn the genes. You have to switch them on and off, as it were, and then measure the metabolism in other areas in plants, and as we do today, also in animals and humans. (Interview 3–1)

Both receive various forms of institutional support and financing. Both grow beyond the feared 'valley of death' between public funding and the first large-scale commercial financing initiatives. Both succeed in bringing new products to the market and commercial success to investors. However, both take markedly different development trajectories.

In the first example – throughout this book referred to as 'Case 3 GENE FUNCTION' a team of researchers at a newly established Max Planck Institute focusing on plant biotechnology search for ways to pinpoint the effect genes have on a cell's metabolism. The rationale of this project is to deliver a radically empiricist response to the question of what genes actually do in cells. Cross-referencing myriad combinations of gene activity with the biochemical complexity of metabolic processes, as well as mapping the pathways between the two is a Herculean task in terms of data handling and combinatorial analysis. The first major obstacle, however, is finding a way to adequately observe metabolic processes. Aided by coincidental encounter, the team identifies gas chromatography as a suitable technology: Metabolic processes leave a mark in a cell's 'exhaust fumes' which can be traced by analysing the spectrum of light reflected by the gas. Gas chromatographers are expensive, but since the institute's initial research capacity is still being built, funds are available. With this important gap closed, the team manages to build an experimental assembly and test relations between gene activity and metabolism in one plant genome. After that, the team leader quickly decides that with this experimental demonstration the job of a basic research organization like Max Planck is done, and from here on more application-oriented actors need to take over.

A small, entrepreneurially minded team is gathered, which takes to advertising the technology to potential investors. A large German chemicals corporation is interested. It is seeking to enter the crop breeding market. Having missed the first wave of gene-based technologies like many other German companies, it is attempting to catch the next. The company sends an experienced executive from their R&D plant in the US as project leader and funds a joint venture with 50 million Marks. He explained:

> And this is a company that did not grow organically, but a company or platform which we built through acquiring and building clusters of excellence, because we, as [company name], entered the market very late. And we said, 'How do we quickly catch up with our competitors, [names of large chemicals companies]?' Yes, by not waiting for things to grow organically, but by approaching the top centres of competence which are already working on the next generation of products. What we do at [company name] and what we also did back then, it to create the genetic and intellectual basis for the creation of these plants of the second and third generation, that is, to optimize; and not the plants you see today: Roundup-ready soybeans; a huge market which is still exploding. But, our products are plants with improved properties, agronomic properties, such as higher yield stability. (Interview 3–1)

At this moment, construction work begins for what will turn out to be the world's biggest and powerful bioinformatics platform of its time and – according to participants – well into the time of data collection for this book. Industrial scale organization and technology – robotics, computers, software, but also recruiting procedures and management structures – are being applied to set up a platform capable of observing the functions of thousands of genes in high throughput mode. The initial aim is to contribute to the mother company's crop breeding activities. Hence, the original experimental technology is scaled-up in every possible way to deliver a form of generalized automated discovery capability. This type of techno-scientific assembly yields economic value if paired with other powerful units – product development, marketing, legal, production, distribution – either via a well-defined market-interface or via integration in one big corporation.

After five years all goals are met: A predefined set of organisms have been tested and patents for several ten thousand genes (according to an interviewed executive) have been submitted and granted. The platform could be taken down, but instead, the mother company decides to keep it and use it as a foundation for future product development activities. Following this decision, around 2005 the company enters into a strategic partnership with a leading American agro-biotechnology corporation. The platform now is to deliver data – and intellectual property (IP) – on gene functions for a whole new generation of products, including genetically modified food crops. The aim is now to replace the first generation of genetically modified organism (GMO) crops with their limited range of functional modifications with newer, more capable varieties in terms of yield, resilience and nutritional value (see Interview 3–1).

At about the same time, another service branch is established. The basic technology's capability extends beyond the realm of crop breeding and includes, for instance, the analysis of human gene functions and cell metabolisms. Hence, it can be used for drug discovery purposes, i.e. the identification of potential targets for new drugs: biochemical 'entry points' which allow intervening in the course of a disease. In its physicality the platform remains the same – a high performance techno-scientific assembly located in one building. But new input and output interfaces are established, both physically and organizationally, and relations to customers, i.e. pharmaceutical companies, are bundled in a service subsidiary.

Thus, in its maturity the technology in question has penetrated the global crop market via the product pipelines of the world's leading agro-food corporations, and at the same time has spread into the pharmaceutical market. Through its materialization as a high-throughput platform, it combines a generic capability with vast economies of scale – not in terms of material production, but in terms of knowledge production. Through its sheer size and its strategic placement within a big corporation striving for dominance in an already oligopolistic market, it is posed to rule, with incremental improvements which are aided by the cross-fertilization between agro-biotechnology and pharmaceuticals, until replaced by another disruptive technology.

In its radicalness, this example is exceptional and yet telling. It features: A new scientific paradigm, materialized in a well-funded institute, pursued by entrepreneurially minded and industry-friendly scientists, matched up with a powerful strategic investor: A leading corporation which is, just at this point in time, in need of a radically new technology. This story may confirm the biotech-critics, who fear excessive knowledge monopolization by big corporations. At the same time, it is full of (seemingly?) coincidental and unlikely encounters, shedding light on the pitfalls encountered by biotechnology innovators.

Let us now consider the second example, or as it will be referred to throughout this book: 'Case 4 BIOMARKER'. In this case, too, the deterministic understanding of the role of genes is revisited. The innovation's key scientific mechanism is DNA methylation: a chemical alteration to the DNA molecules, which is relevant for their activity, that is, whether they are 'switched on' or 'off'. The observation of methylation patterns is essential to the field of epigenetics, which studies the ways in which gene activity is regulated. Epigenetics is, at the beginning of this innovation process, a young field and a young community. It is meeting resistance from more orthodox geneticists, particularly in Germany, as one scientist involved in the innovation recalls:

> As an epigeneticist, you don't have a very good standing in Germany anyway. You had to run against a front of traditionalists very much, and they made your life a misery. [...] Within biology, but mostly within genetics. They said, 'That guy somehow does some descriptive stuff. Interesting, but balderdash'. That was pretty hard. (Interview 4–6)

The scientist also highlights another feature of epigenetics as an epistemic endeavour:

> And above all, in the beginning in epigenomics, 'green' and 'red' biologists talked to each other a lot, and they still do. This is very rare. It is very rare that concepts, ideas and implementations of ideas are exchanged between green biotechnology or biology and red biology. This takes place here and leads to mutual stimulation, because these are universal concepts. (Interview 4–6)

At the Max Planck Institute in question, scientists are developing the mathematical and computational prerequisites for a reliable and efficient measuring and testing of methylation patterns in small samples. In this young field, cancer diagnostics, and potentially therapy, appear as the 'natural' field to apply new developments in epigenetics. Cancer cells deviate from healthy cells regarding their DNA methylation patterns. A team of seasoned clinicians and experienced entrepreneurs in Seattle is, simultaneously to the work undertaken at the MPI in Berlin, in the process of laying the scientific – epigenetic – foundations for a new cancer diagnostics company. This company would later play an important role in the innovation process.

A young PhD student employed in the Berlin group suggests submitting a patent for the methylation technology and writes a business plan with the aim to launch a start-up. He receives support from his superiors and rallies both to build a founding team – all young PhDs – and to enrol investors. The founders aim for venture capital (VC) and entrepreneurial independence. At this point the business case is vague: It can entail a generic technology platform, a diagnostic product or, theoretically, even a new drug. But in Germany's VC scene – which for the first time is truly thriving – the technology's great potential is seen as outweighing its uncertain application. A Munich-based VC becomes lead investor and quickly assembles an international consortium involving British and American VCs.

The start-up is accommodated in an unused inner-city rear courtyard building. The entrepreneurs themselves transform the building into lab space in handicraft. Equipment is partially created through DIY and improvisation. PhD students from the 'mother' institute are employed as scientists in a joint framework with the institute. Commune-style, night-long discussions produce ever bolder technological and scientific visions. While the vibrancy and hands-on nature of events resembles those just a few kilometres away (Case 3), the absence of industrial structure and leadership is palpable. Yet, the VC consortium drives for professionalization: The aforementioned cancer diagnostics start-up in Seattle becomes the partner in a merger – or rather a take-over, following the unlikely situation that the German start-up is well funded, and the American is not. The investors want a sound business case and welcome the Seattle team's focus on diagnostic products.

In addition to this 'injection' of experience and professionalism, the new company enters a strategic collaboration with a Swiss pharmaceutical and diagnostics corporation who acts as a pioneering user. The new technology is to be tested and incorporated into a new series of diagnostic products: blood-based tests for different types of cancer. While the new partner provides all infrastructures and competences necessary for large-scale testing, the founders can concentrate on technology development. During this strategic collaboration, in 2004, the start-up goes public successfully – to this day an unusual accomplishment for a German biotechnology company. Shortly after, the collaboration is broken up due to 'strategic differences'. The original founder, who has driven the enterprise forward to this point, leaves in anger shortly after. The start-up, now an independent, publicly listed company with two sites in Berlin and Seattle, decides to go for diagnostic product development, capitalizing on the investment already made by the former partner and a favourable IP situation.

In the following years, the company transforms dramatically. After going through a leadership crisis caused by the abrupt ending of its key strategic collaboration, the company starts adopting the industrial style of working that was absent before. New, less exclusive partnerships with four international diagnostics companies are established to achieve better market penetration. Europe and America differ strongly with regard to their market-entry conditions for diagnostic products. As respondents explained, the American regulation agency, Food and Drugs Administration (FDA) requires clinical trials to prove the effectiveness of a diagnostic test. In Europe, the industry is more self-regulated and freer to set its own goals, which then provide the reference point for certification. Other markets have other rules. More importantly, diagnostics is a business strongly influenced by national working cultures and practices, particularly in diagnostic laboratories, which essentially are the customers of a diagnostic product. Therefore, a lot of work is put into building networks of intermediaries and partners in prospective market countries. The Seattle site becomes particularly important for setting up clinical trials in the US as well as lobbying the FDA. In 2011 the first blood-based colon cancer test receives regulatory clearance in the US and Europe and enters the market. Colon cancer as a very common disease in

Western countries, which also spend much on its early detection, is the logical choice for a lead product. A second product for lung cancer follows several years later.

In many ways, this case tells a similar story as the one presented before: The role of entrepreneurial thinking *within basic research*, the importance of large corporations who act as strategic partners, the opening of opportunity windows in Germany in the late 1990s, the drastic shift in working routines halfway through the innovation process, the dynamics of transnational integration. And yet, here is a case with a decisive influence of venture capital as opposed to purely industrial investors. Partial failure, back and forth are more prominent in this case. And finally, rather than turning a new scientific paradigm into a high-performance platform combined with product development, a different path unfolds. It leads to a small selection of diagnostic products. An independent, medium-sized, publicly listed company is created, adding to the comparatively small stock of such in Berlin given the city's size. In addition, the innovation led to the establishment of an entirely new market segment: blood-based cancer tests. Why cancer diagnostics and not another field, given that the basic science is rather general? Asked more generally: How is it possible, as happens very often, that two innovations with very similar starting conditions take very different trajectories? There is ample reason to suspect that the interaction between scientists and investors, capital and knowledge creation, exert a strong influence on the path of an unfolding innovation.

This book is about the relationship between capital investment and knowledge creation, and the influence this interplay exerts on the paths of unfolding innovations which lead to economic value creation. Thus, inevitably, it is about the relationship between investors and their agency, and, on the other side, entrepreneurs and scientists as well as their agency. I look onto this interaction from a particular perspective: the economic geography of innovation processes. Economic geography has a natural interest in innovation. Territorial innovation systems, both national and regional, have been popular and extensively researched concepts for some time, because they underscore the diversity of territorially grounded institutions and inter-institutional arrangements – and with them the innovation outcomes they produce (Moulaert and Sekia 2003; Morgan 2004). However, this work seeks to contribute to a more recent process-oriented literature in economic geography (Crevoisier 2014; Ibert and Müller 2015) rooted in relational economic geography (Bathelt and Glückler 2003). It centres on the unfolding of innovations through time and space: across multiple distances and proximities, places and territories, contexts and situations. This literature seeks to explore the role of mobility and difference in innovations. It draws on ideas of relationality as opposed to more structural notions of systems and institutions. Relational economic geography is not without criticism: Some see its lack of a big theory as an indication of analytical weakness (Yeung 2005). I wish to show that a relational economic geography of innovation processes has a particular potential to provide an integrated view on the techno-scientific and the financial value-related aspect of biotechnology innovation. I further seek to make conceptual contributions to the field which help to better realize this potential.

The first major contribution I wish to make in this book is to highlight the relational work of building, maintaining and ending relationships in innovation processes, of applying particular ideas of value or proper practice and of changing those that have been applied previously. On a more abstract level, this work is also about the agency of creating and recreating particular ideas of relatedness and non-relatedness, proximity and distance, which are sometimes taken for granted.

The second contribution is the systematic integration of investors and investment practice into conceptualizations of innovation processes. There is an extensive literature on the geography of venture capitalism (Chen et al. 2010) focused either on the regional or the firm-level, and this literature cannot be ignored in an endeavour such as this. Also, the wider field of financial geography, which has surged in recent years, holds relevant insights for this work (e.g. Pike and Pollard 2010). However, integrating investor agency and investment practice into an innovation process perspective – *and venture capital, although strongly associated with biotechnology, is not necessarily the most important type of investment here either qualitatively or quantitatively* – is still worthwhile or can provide novel insights. Notably, as will be seen, biotechnology innovations both precede and succeed biotechnology companies (VC funded and otherwise) which often have the character of temporary organizational platforms. I attempt to study the valuation of knowledge in investment on one level with other rationales in innovation processes, and at the same time remain open about the nature and possible diversity of investment rationales. Investors, too, participate in the relational work of building proximities and distances, of initiating and ending relationships and thus contributing to particular innovation outcomes.

The practical relevance of this contribution is that it chimes in with a recent wave of research-based innovation policy concepts which place agency and leadership before structures or what might be called 'thickness-building', and knowledge work across distance before purely territorially bounded or proximate relationships. So far, venture capital policy tends to be among the most territorialized and proximity-centric of all innovation policy fields.

The innovation histories presented in the beginning represent two of eight qualitative case studies, each one covering an innovation process in German biotechnology. In each case study I try to reconstruct ex-post the relational work of creating, changing and ending the socio-material relations which make an innovation process advance – including investment relations. The case studies cover innovation processes in the time-span from roughly 1990 to 2015. There is considerable temporal overlap between them. The cases are not strictly studied from a comparative perspective. Instead, they also provide glimpses on a highly dynamic ecology. Their unfolding reflects (and in a smaller way enacts) bigger changes.

Throughout the next sections of this introductory chapter I will lay out a framework of analysis. Section 1.2 will discuss biotechnology from the perspective of a fundamental tension – between a political economic and a techno-scientific view – which can serve to inform advances in relational economic geography. Section 1.3 will present the key literature used in this study: the relational economic geography of innovation processes. Sections 1.4 and 1.5 will discuss key areas in which I seek to expand the aforementioned literature, namely the agency of making and unmaking relationships in innovation processes and the specificities of venture capital investment relationships. Finally, in Section 1.6, I will discuss the methods used in this study. In Chapter 2 the specificities of biotechnology in Germany will be presented. Chapter 3 will focus on the dynamics of creating and unfolding innovative relationships in early stages of biotechnology innovation, including the role of investors. Chapter 4 will focus on later stages and the inevitable socio-material transformation each (successful) innovation must undergo. Likewise, the role of investors and investment will be discussed. In Chapter 5 I will draw conclusions.

1.2 Biotechnology and finance – between technoscience and political economy

According to Birch (2016) there is a tension between a 'technoscientific' view on biotechnology, rooted in the Science and Technology Studies (STS), and a political economic view: The earlier focuses on the 'commodification' of knowledge (the creation of economic goods out of biological knowledge) and assumes that economic value is somehow latently inherent in organic processes and matter. The latter focuses on the practices and agencies of economic actors who evaluate and commercialize biotechnological knowledge in a way specific for contemporary capitalism. Crucially, in Birch's own political economic view, biotechnology is not so much valued for concrete products which emerge from scientific knowledge. Instead, the dominant form of economic value creation in biotechnology is speculative in nature. The biotechnology firm is a 'financial artefact' whose value is constructed by actors in the sphere of finance (e.g. venture capitalists), and which is in itself a tradable commodity. Connections to the promise of actual use value (as in the case of a new cancer therapy) are important as legitimizing narratives, but not much more. The evidence to date seems to confirm the thesis that in biotechnology, financial valuation (extremely high) and real outcomes in terms of new usable products (very low) are almost entirely disconnected (Mirowski 2012; Pisano 2006).

Relational economic geography has the potential to build bridges between the two perspectives. Therefore, I will give a short introduction to biotechnology with the indicated tension in mind. I will also problematize some limitations of the political economic view when it comes to the study of *innovation* (i.e. the creation of real and novel products) *under conditions* of 'financialization' and 'assetization' (Birch 2016).

A closer look at *bio-technology* reveals that even with a focus on tech-noscience, the connection to organization, industry and politics is never far. According to the OECD, the term biotechnology denotes 'the application of science and technology to living organisms, as well as parts, products and models thereof, to alter living or non-living materials for the production of knowledge, goods and services' (OECD 2005, p. 9, quoted in Nolting and Mietzner 2010, p. 10). Modern biotechnology as an industry and as a tech-nological paradigm is strongly associated with recombinant DNA: the ability to isolate gene sequences, extract them from living organisms and induce them into other organisms. The technique is also referred to as genetic engi-neering. The respective technological breakthroughs occurred in the 1970s and have a more direct linkage to the emergent biotechnology industry than purely scientific discoveries such as the Watson Crick model in 1953. OECD provides a more comprehensive list of modern biotechnologies. It includes DNA/RNA (Ribonucleic acid) technologies, protein and other molecule technologies (such as engineering and synthesis), cell and tissue culture and engineering, process biotechnology techniques (such as fermentation in bior-eactors), gene and RNA vectors (using viruses to carry genetic material into cells, e.g. for therapeutic purposes), bioinformatics and nanobiotechnology (Nolting and Mietzner 2010; OECD 2005).

This difference is best illustrated by considering the difference between genetics and genomics. The former is an epistemology aimed at the illumina-tion of the mechanisms of inheritance. The latter is a scalable technical pro-cess of sequencing, mapping and cataloguing, and also a potential business model (Sunder Rajan 2006). Toward the end of the 1980s, the United States government, specifically the department of energy, initiated an international endeavour to sequence and map the entire human genome. The resulting Human Genome Project ran from 1990 to 2003, involving scientists from six countries. Initially, scientists rejected the undertaking, because it was not hypothesis-driven. The genomic and potentially commercial nature of large-scale gene sequencing was further elucidated by the competitive dynamics which commenced shortly after the start of the official, public project. Using advanced polymerase chain reaction (PCR) robots, entrepreneur Craig Venter and his company Celera Genomics challenged the HTG consortium to a race for the completion of the sequencing process. In this race the future role of automation and computational biology became visible in a dramatic way. The robots were provided by a company called Applied Biosystems (ABI), which gained drastically in reputation and market power in the process (ibid.).

These expectations were disappointed. But the study of innovations in bio-technology can benefit from an appreciation of the basic structure of this and other large-scale genomic endeavours. An organic mechanism, a) observed and experimentally reproduced in a field of modern biology (in this case gene sequencing), is b) taken out of its epistemic context by actors like govern-ments and corporations and placed in an environment of factual or envi-sioned application. In this environment, it is scaled up. This transformation requires c) a scalable enabling technology. Thus, progress or innovation cannot be conceptualized as epistemic advancement alone. Instead, the meeting in time and space of epistemic advances, enabling technologies and an application context define an innovation in biotechnology. The fields of application are a very prominent basis for the differentiation of biotechnology: 'red', i.e. medical biotechnology is separated from 'green', i.e. agricultural biotechnology. Other denominations like 'white' (chemical/industrial biotechnology) and even 'blue' (maritime biotechnology) exist. These differentiations can coincide with epistemic differentiations or the enabling technologies used, but crucially, they do not necessarily do so.

Long before the Human Genome Project, the foundations for biotechnology were laid – along with shifts in policy which have widely been labelled as 'neoliberal' (Kotz 2011; Nölke 2009), such as liberalizations of capital markets and the strengthening of intellectual property rights (IPR). But public institutions were equally important. In biotechnology, the state is a leading actor (Mazzucato 2013; Prevezer 2008). Several authors stress the importance of publicly funded life science research in the United States, specifically through the National Institutes of Health (NIH). NIH research funding, both through grants and directed at research in the institutes themselves, is perceived as the key resource for the creation for the scientific basis of biotechnology. Since the Second World War, health research, specifically cancer research has been a priority of public policy in the US. The NIH system represents the most influential institutional form in which this research was organized (Bartholomew 1997; Giesecke 2000). Early biotechnology was not addressed as a purpose in itself but as a means to an end: improving public health.

The growth of venture capitalism (VC) in the 1980s is maybe the development most strongly associated with the emergence of a biotechnology industry. In fact venture capitalism emerged in the United States in the 1940s (Gompers and Lerner 2001). Its breakthrough came in 1979 when the Employee Retirement Income Security Act (ERISA), which severely limited pension funds' freedom to invest in high-risk asset classes, was changed. During the 1980s, public and private pension funds led an increasing influx of funds into venture capital funds, i.e. funds which invested in high-risk ventures seeking a profitable sale on the stock market or initial public offering (IPO). In this period, a fundamental change in the industrial landscape took place. A radically new high-tech industry emerged, including modern information technology and biotechnology. Stock market value and employment shifted drastically from large established corporations to young technology enterprises. In the case of biotechnology, few newly founded but dynamically growing companies such as Genentech defined and configured the field. These pioneering companies were venture-capital funded (Gompers 1994). In the 1990s and early 2000s, despite a severe financial crisis, venture capital was seen as the driving force behind each surge of technological innovation, specifically the internet economy (Zook 2002, 2004). According to Florida and Kenney (2000) venture capitalists established a new innovation model by cultivating networks of universities, financial institutions, large companies and start-ups, thus overcoming the barriers of both individual entrepreneurship and corporate innovation.

The emergence of VC is often placed in the context of a broader development labelled financialization, which extends far beyond the financing of entrepreneurship. Financialization denotes 'a pattern of accumulation in which profits accrue primarily through financial channels rather than through trade and commodity production' (Krippner 2005, p. 174, cited in Birch 2016, p. 464). It leads to an increasing size and importance of the financial industry and with an increasing importance of financial valuation practices in economic life (Pike and Pollard 2010). The first aspect includes the emergence of new financial actors and new financial intermediaries, such as hedge funds and large-scale institutional investors outside the traditional banking sector. A more recent development in this area is the rise of a dedicated securitization industry, which creates ever more complex and multi-layered synthetic financial products. Private households were enrolled in such financial circuits in new ways, both as debtors and as creditors, adding weight to the system and intensifying dependencies (French, Leyshon and Wainwright 2011; Sokol 2013). In addition to the financial industry, the degree to which other industries engage in financial investment and speculation adds to the growing quantitative relevance of finance. This combination of growth and internal complexity – at the expense of other economic activities and partly as a response to trade imbalances to the disadvantage of the United States – is seen as a systemic source of volatility. The finance-dominated accumulation regime is thus structurally characterized by repeated crises (Sokol 2013; Stockhammer 2009).

The second aspect relates to the metrics of value which are employed in economic practice. 'Shareholder value', the prioritization of shareholder interests, has fundamentally transformed the ways in which stock-market-listed companies operate. Critical scholars argue that the ability to engage in long-term research and development activities was lost due to the emphasis on quarterly profit figures (French, Leyshon and Wainwright 2011; van der Zwan 2014). Revenues were more likely to be handed to shareholders as dividends than to be invested in productive or knowledge-intensive capabilities. Shareholder-value-based corporate governance has grown beyond the US/UK context and become an international phenomenon.

Another factor for the early growth of US biotechnology is seen in the large size of the American market for health products, both with regard to people and to purchasing power. Through its competitive private insurance market with no limitations on prices, the United States acted as a lead market, which fostered biotechnology innovation through demand-pull. The liberal capital market with large-scale private pension funds and a stock-enthusiastic public provided the needed investment capital. In the capital market, both investments prior to an IPO, hence, venture capital, and investments on the stock market contributed to the young industry's growth (Prevezer 2001).

Changes in the institutional regulation of IPR likewise encouraged entrepreneurial activity. In the 1980 Patent and Trademark Law Amendments Act, more widely known as the Bayh-Dole Act (Popp Berman 2008; Zeller 2008), universities are granted ownership of all IP emerging from scientific research conducted by their personnel. This change favoured the creation of professional IP management structures, relieving scientists of the effort and cost of patenting and patent enforcement and increasing the likelihood of commercialization. Comparing European and American property rights regulations, Giesecke (2000) also comes to the conclusion that US patent legislation is more compatible with academic entrepreneurialism, since publication of information prior to patenting does not render a patent application invalid ('first to invent' vs. 'first to file').

The importance of IP rests in the central role codified information plays in biotechnology. Sunder Rajan (2006) argues that at the core of biotechnology's accessibility to financial valuation is codification. In interaction with biological matter, information is created. This information, stored in databases which act as intermediaries, is used to develop practical, epistemic knowledge, which is then reapplied to the biological matter (see also Latour 1999). Hence, in its essence, biotechnology is an information technology. Information can be owned and information ownership is at the heart of commercialization activities in biotechnology. A focus on codified information is very compatible with the American brand of financial capitalism. Casper and Kettler (2001), for instance, highlight the important role of codification and quantification in American management practices, based on an institutional environment which favours arm's-length, short-term relationships between companies and their employees.

By focusing on the nexus of biotechnology firms, investors and the stock market, it is easy to arrive at the conclusion that biotechnology is primarily a speculative field with a few rentiers in the centre. Andersson et al. (2010) refer to biopharmaceuticals as a 'speculative innovation, capital market liquidity business model', which is fundamentally different from other business models. Biotechnology firms are usually created as university start-ups and undergo a succession of investment by different actors (business angels, venture capitalists, pharmaceutical companies, stockholders) until they establish products on end-user markets. Consequently, they are constructed as elements of investment portfolios, in which product pipelines are quantifiable assets. Crucially, no real certainty exists regarding development progress in development pipelines and the risk of failure is tremendously high. The reason is a 'profound and persistent uncertainty rooted in a limited knowledge of human biological systems and processes' (ibid.). Therefore, complementary narratives of progress and shareholder value are an important currency in the construction of value. The narrative of the 'blockbuster drug' is particularly powerful in fuelling stockholder fantasies (Montalban and Sakinc 2013), but also leads to concentrations of investment on the largest, most profitable markets. Other ingredients are capital market liquidity and risk appetite on the side of investors, specifically on the stock market.

Lazonick and Tulum (2011) specifically highlight the role of a speculative stock market. According to them, biopharmaceutical product development is such a long-term process (sometimes 20 years) and riddled with such high risk, that the financing of the US biotechnology industry relied and still relies on one key factor: the readiness of stock market investors to buy shares and support IPOs of companies, which are far from having an end-user product. Thus, in a risk-friendly environment, venture capital investors and even pharmaceutical corporations can realize a profit from investing in biotechnology even in the absence of a marketable product. Lazonick and Tulum raise the question of whether this system is capable of creating effective and affordable drugs on the long run, especially in the face of repeated financial crises. This line of argument strongly resembles Zeller's (2008) account, in which organized investors in biotechnology are, first and foremost, rentiers.

Critical scholars see both the institutional changes to university patenting (Bayh Dole Act) and the drastically increased numbers of filed patents since the early 1980s, primarily held by corporations, as a campaign of dispossession (Zeller 2008). Scientists are deprived of the results of their work. Public research organizations (specifically the NIH system) are likewise dispossessed. This dispossession is understood as the more civilized variant of an essentially violent mechanism, which, in the opposite extreme, deprives rural farming communities in the global south of their knowledge-cultivation capabilities. The resulting IP monopolies allow corporations to extract rents even more effectively than earlier capitalist rentiers could through their ownership of machinery. Every 'commercial' user – and this includes diagnostic departments in hospitals – has to pay royalties to the patent holder. This can make the widespread usage of new technologies prohibitively expensive. The problem is aggravated by frequent extensions of the usual patent duration based on technical and legal arguments (Zeller 2008).

While capitalists appear as rentiers within the economy, the USA appears as a biotechnology rentier on the global scale (Zeller 2003). Since the 1980s, the United States has been a net capital importer, attracting funds, among others, from Europe and oil-wealthy Gulf countries. In this account, biotechnological industry concentration reflects capital concentration. As Cooper (2007) puts it:

> Since [1980], the life sciences have played a commanding role in America's strategies of economic and imperialist self-reinvention. Over the past few decades the U.S. government has been at the center of efforts to reorganize global trade rules and intellectual property laws along lines that would favor its own drug, agribusiness, and biotech industries. Moreover, the unique position of the United States itself in relationship to world financial flows has meant that even the most speculative of its life science enterprises has attracted a constant, and incomparable, flow of funds. (Cooper 2007, p. 4)

The critical and the more optimistic account of biotechnology may be reconciled by turning to a less structural type of argument. Sunder Rajan (2006) points out that there was no predetermination in the unfolding development path of the American biotechnology industry. While the corporatization of knowledge was, on the one hand, hegemonic and rapid, it was, on the other hand, also contingent and contested. The places, situations, actors and socio-material practices of commercialization warrant attention, rather than structural presumptions and normative judgments. More recent studies of financialization build on this premise and adopt a more relational view. They find that financialization can indeed lead to an uprooting of localized knowledge and economic practices and their replacement by standardization, ownership concentration and an appropriation of formerly intangible values – like local identity and image (Pike and Pollard 2010). However, they also find that an opening of economic sectors and regional economies to financial valuation can infuse new life and lead to a renaissance of entrepreneurial energy in encrusted business relations (Zademach 2009). Viewing biotechnology from a more 'ecological' and embedded perspective and refraining from over-emphasizing the firm-level may provide a more balanced perspective.

Biotechnology companies occupy positions between academia and large-scale industries. Only a small number of biotechnology companies, particularly in the United States, have grown to the size and importance of global players. The most cited examples are the biopharmaceutical companies Genentech and Amgen. Genentech, founded in 1976 as a drug development company based on recombinant DNA technologies is understood as a model, which configured the organizational field of biotechnology (Andersson et al. 2010; Powell and Sandholtz 2012). Most biotechnology companies stay close to their origins as start-ups from universities and other research facilities. In fact, biotechnology start-up companies are understood as the most effective mechanism of transferring knowledge from an epistemic into an application environment. The ability to create a start-up company in an academic environment is strongly affected by academic career structures and incentives. According to Prevezer (2001), creating a company while remaining in academia or at least keeping in close contact with research institutes is comparatively easy in the American institutional system. Academics are not discouraged from making entrepreneurial choices, and universities perceive engagement with the start-up scene as mutually productive.

Generally, biotechnology is a dynamic knowledge economy in which dedicated (or core) biotechnology companies interact with research institutes, industrial corporations and specialized service and technology providers. According to Powell and Sandholtz, biotechnology companies and entrepreneurs resemble amphibians, because they operate in the academic and the industrial realm (Powell and Sandholtz 2012). Institutional boundary spanning and openness are both preconditions for the emergence of the new organizational form of the science-driven biotechnology company and elements of the day-to-day operations of such companies. This understanding of biotechnology as a cross-organizational knowledge ecology is reflected in the empirical focus, particularly in economic geography, on regional concentrations (clusters) of biotechnology companies. In the United States, the preeminent clusters are the San Francisco Bay Area and the Boston Metropolitan Area. These regional concentrations are understood as birthplaces of biotechnology both as a technological paradigm and an organizational form.

Outside the US, too, biotechnology is highly concentrated in a small number of well-established high-tech regions (Cooke 2002; Gilding 2008). A considerable amount of scholarly energy is invested in analysing the different structures of regional networks, the knowledge bases and specializations, the institutional forms of technology transfer and the variegated actor constellations in such regional clusters and innovation systems (Coenen, Moodysson and Asheim 2004; Coenen et al. 2006). Regional concentrations of biotechnology are subject to considerable structural inertia or even path dependence. On the one hand, their emergence is contingent on pre-existing capabilities and awareness (Feldman and Francis 2003); on the other hand, they are reinforced by powerful actors (like pharmaceutical corporations) who establish presence in them in order to benefit from regional knowledge dynamics (Cooke 2002).

Biotechnology companies usually do not have access to end-user markets. These markets are dominated by large corporations, particularly in the pharmaceutical sector (Montalban and Sakinc 2013). Pharmaceutical companies began adapting to the biotechnological start-up dynamic in two ways. They began restructuring their own internal organization in order to mimic biotech start-ups, for example by establishing cross-departmental project teams and internal competition (Zeller 2002). And they reduced internal research efforts in favour of tapping into the start-up ecology and thus employing a more evolutionary approach to new product development (Andersson et al. 2010; Mittra 2007; Montalban and Sakinc 2013). Strategic relationships were formed, which granted biotechnology companies access to markets and pharmaceutical companies access to new technologies and thus a rejuvenation of their product pipelines. From the perspective of pharmaceutical corporations, biotechnology yields the promise for a new generation of patentable drugs. The number of market admissions of new pharmaceutical substances has been declining since the 1970s. A substantial number of 'blockbuster' type drugs was about to reach the end of their 20-year patent protection around the year 2010. In this 'patent cliff' situation, engaging with biotechnology has become an existential matter for many pharmaceutical companies (DeRuiter and Holston 2012). The promise of a biotechnology revolution has, however, so far not materialized.

The agency to be considered asking for the realization and non-realization of innovation is distributed across many actors. Ideas from political economy have their place alongside ideas from traditions like STS.

1.3 Relational economic geographies of innovation processes

This study focuses on innovation, i.e. the creation of new products ('commodification' of knowledge) rather than 'assetization' or 'financialization' (Birch 2016). The latter are, however, important conditions for contemporary innovation dynamics and require careful consideration. A classical definition of innovation is the implementation of new combinations (Schumpeter 1997 [1911]) to solve a practical problem. It comprises the introduction of novel solutions to practice, in the case of economic innovations via the market, as achieved by combining hitherto unconnected elements. Following this definition, innovation is not identical to invention. While invention is the creation of an innovative idea, innovation also includes all steps necessary to implement a novel solution in economic practice. These steps, too, can be conceptualized as knowledge-creation, as they involve the generation of new routines, roles and practices. This understanding, however, is at odds with the way knowledge has long been conceptualized in regional economics and economic geography (Ibert and Kujath 2011). Cook and Brown call this traditional way the 'epistemology of possession' and state that it is 'essentialist'; in it, knowledge is conceptualized as something objective and immaterial, and furthermore, as a good which can be possessed (Cook and Brown 1999). On this basis, it is difficult to grasp the dynamics of innovation fully. The reason is that this epistemology would prioritize the act of discovering or inventing, which would supposedly lead to genuine and objective novelty when compared with the activities 'merely' directed at implementation. This does not appropriately reflect the reality of innovation, which is in fact characterized by a multitude of existential challenges during implementation.

In economic geographies of innovation, the essentialist perspective on knowledge is increasingly replaced by a concept of 'knowing in practice' (Ibert 2007) or a 'pragmatist' epistemology of knowledge (Cook and Brown 1999; Dewey 1933). It stresses that knowledge is best understood as a prerequisite for, and an element of, action, and that it lives in situated, material and to a certain degree, embodied practices. The 'practice turn' or 'practice shift' (Jones 2013) has affected many research fields in the social sciences. For some scholars, practices are a phenomenon very similar to routines. According to Jones and Murphy, socioeconomic practices are 'stabilized, routinized or improvised social actions that constitute and reproduce economic space, and through and within which diverse actors [...] and communities organise materials, produce, consume, and/or derive meaning from the economic world' (Jones and Murphy 2011, p. 367). Jarzabkowski, Matthiesen and van de Ven (2009), in their definition, highlight the importance of practices' embeddedness in material environments: 'A practice approach examines how actors interact with, construct, and draw upon the social and physical features of context in the everyday activities that constitute practice' (p. 288). While the essentialist epistemology of knowledge adheres to a distinction between implicit, 'tacit' knowledge on the one hand and explicit, codified knowledge on the other, a practice-based approach treats all aspects of knowledge-creation, including explication and interpretation of codified knowledge representations, as situated and material. With this shift, the contingencies, context specificities, challenges and serendipitous elements in innovation become more accessible conceptually.

In addition to the practice perspective, relational economic geography highlights the relational nature of knowledge and resources (Bathelt and Glückler 2003, 2005). According to this view, a competitive advantage is not the result of accumulation or stockpiling of 'objective' knowledge, but of a temporary knowledge gap between actors, which can be created and exploited. Knowledge is created and cultivated in social constellations. While the epistemology of possession focuses on the firm as the social entity cultivating knowledge, the practice perspective adds an emphasis on the community as a social nexus of knowledge-cultivation (Amin and Cohendet 2004; Brown and Duguid 2001). 'Communities of practice are groups of people who share a concern, a set of problems, or passion about a topic, and who deepen their knowledge and expertise in this area by interacting on an ongoing basis' (Wenger, McDermott and Snyder 2002, p. 4). Consequently, learning is closely linked to social participation. More broadly, the creation of knowledge can be understood as a process of positioning oneself in a relational space. An innovation thus takes place when actors set themselves apart from others by establishing a novel or different understanding of a particular task (and associated practices), which yields potential improvements (knowledge advantage), and by materializing it in structured, routinized action.

However, situated material interaction is not necessarily routinized. It can also be highly creative, providing starting points for innovation. For example, for communities of practice in architecture, embodied experiences with models and material samples in a studio setting – sharing the tactile experience, discussing models and viewing them from different angles – is an essential part of design work (Faulconbridge 2010). Thus, learning is based on physical interactions and kinaesthetic experience. The role of objects goes beyond that of tools in such learning dynamics. Concepts like 'dynamic affordance' and 'productive inquiry' denote the continuous creation of learning opportunities if human and non-human bodies interact over time (Cook and Brown 1999).

Intuition, affect and emotion are also relevant factors (Ettlinger 2004). Using the emergence of the fingerboarding community as a case, Brinks and Ibert (2015) show how people interact with objects in an open, playful form, how they are emotionally affected by objects without assigning them specific purposes but sensing a potential. Brinks and Ibert refer to this process as 'tinkering' and describe it as the open search process which leads to a new kind of practice and ultimately new commercial products. This idea that the identities of objects in situations of knowledge creation are not necessarily determined but open is reflected in the concept of the 'epistemic object': it is characterized by its openness and 'non-identity' with itself, i.e. its permanently changing purposes and boundaries (Ibert 2007; Knorr Cetina 2001; Miettinen and Virkkunen 2005) The epistemic object is a notion specifically used to describe processes of scientific knowledge-creation. A more general formulation would be that in dynamics of knowledge-creation, the identities of human and non-human actors as well as their relationships are continuously explored and negotiated.

Several concepts exist to describe the organization of innovations in time. In this discussion, a tension exists between accounts of innovation as a linear, path-dependent process (Pavitt 2005) and innovation as a feedbacked, 'chain-linked', 'permanently beta' dynamic of continuous adaptation (Kline and Rosenberg 1986; Neff and Stark 2004). In recent years, user-induced, 'open' and community-based innovation has received attention, adding weight to views which reject linearity and linear-innovation models. This discussion reflects a trend in conceptual thinking and arguably a real-world trend towards non-linear, user-driven innovation (see Franke and Shah 2003; von Hippel 2005). Theorizing innovation processes in an open way, i.e. without making any ex ante statements about their linear or non-linear nature, remains a challenge. Recently, however, the process character of innovation has received increased attention in economic geography. Innovation biographies have been established as a methodology to study the emergence and time-spatial unfolding of innovations (Butzin and Widmaier 2016). This new approach is in the process of evolving from merely a new form of gathering data on innovation, towards a new paradigm in economic geography in dealing with the time-spatial patterns of knowledge-creation.

Drawing on the concept of the product ethnography and its research imperative 'follow the thing' (Cook and Harrison 2007), Ibert, Müller and Stein (2014) conceptualize innovation processes as emergent, idea-centred knowledge networks, which can be traced by 'following the idea'. Each innovation studied is ascribed a characteristic 'innovative idea', which, from conception to implementation as a new economic practice, materializes in multiple knowledge practices and contexts. The approach takes a linear view of innovation processes as a matter of conceptual necessity: If an innovation, defined by a concrete shift in application practice, is identified ex post and then analytically reconstructed, its course is linear. But linearity also carries meaning beyond this obvious notion. The cases studied show that innovation processes display properties of path-dependence and one-shot operations (Rittel and Webber 1973; see also Balconi, Brusoni and Orsenigo 2010). Innovations frequently involve open search processes, during which innovators do not know what they are looking for until they find it (Stark 2009). Once they have found it, the cognitive framing of the search situation is radically transformed. At this point, there is no going back to the previous 'innocent' state.

Van de Ven et al.'s (1999) seminal work on the 'innovation journey', which is based on more than a decade of intensive case study work, provides an important reference point for this work. In incredible depth and detail, the authors reconstruct the multitude of individual agencies and roles, relationships, organizational forms and interactions occurring in innovation processes. Rather than being entirely random or entirely linear, innovation processes are *complex*, containing elements of *chaos* (in the sense of non-linear causation) the authors find. Accordingly, this account of innovations reflects a plethora of possible influences. This complexity, however, is also a disadvantage: In all the complexity, it is difficult to identify the common lines and patterns (Edmondson 2000). The work is also largely a-spatial.

Ibert, Müller and Stein (2014) structure innovation processes using an inductively developed, generic phase model. One of the model's main features is that it treats unintended, undirected search prior to an actual problem definition – an accomplishment which is often innovative in itself – as a distinct phase in innovation processes. Such a conceptual integration of open search accounts for the often unintended and unpredictable nature of innovation, and also builds a link to ideas of user and application-driven innovation. Secondly, the (initial) validation, i.e. the first materialization of an idea, is conceptualized as a phase of experimentation, which can take place in a science laboratory, but also in any other social context. Thirdly, it distinguishes a local (both in the physical and the relational sense) validation of ideas from a later wider, more robust validity as a precondition for an idea's general application. This mobilization process, too, can take place in a variety of social contexts (Figure 1.1).

The model is not biased in favour of scientific or technological modes of knowledge-creation. This means that very different types of innovation – classical, science-driven innovation as well as 'permanently beta'-style, user-driven innovation – can be described and analysed with the phase model. Due to this openness, the model serves as a bridge between a variety of approaches: economic and technological innovation models on the one hand, more general descriptions of societal change, institutionalization (DiMaggio and Powell 1991) and legitimization (Johnson, Dowd and Ridgeway 2006) on the other. The latter two theorize the introduction, the spread and finally the general application of new social practices. Johnson, Dowd and Ridgeway (2006) explicitly distinguish local and general validity.

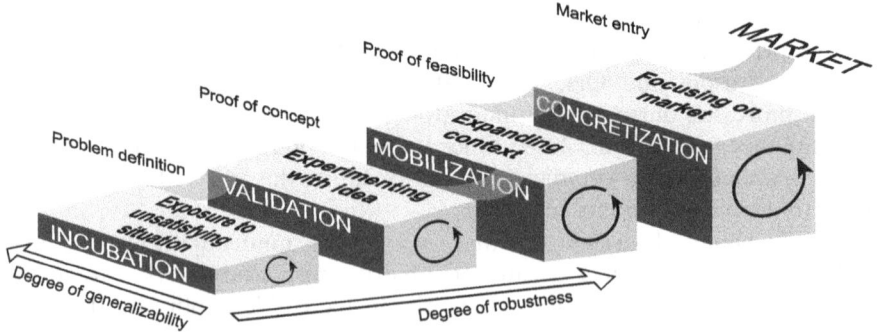

Figure 1.1 Phase model of innovation processes
Source: Ibert and Müller 2015

In order to make ideas valid in an environment larger than the original context in which they emerge, knowledge needs to be made mobile. Understanding these processes is a challenge in a 'knowing in practice' based framework: the practice perspective explains first and foremost, how knowing and learning are place-bound, materially situated and dependent on social relations which operate on a functional as well as a non-functional (i.e. personal, emotional) level (Lorenz 2001). It can explain why it is tremendously difficult to transfer good practices from one site to another (see Szulanski 2003), but it is weaker in explaining why this happens, nevertheless. Therefore, an understanding of knowledge as an exclusively locally situated phenomenon has come under critical scrutiny, especially from scholars of entrepreneurialism and innovation.

One way to expand the perspective is to assume a world of many locally situated communities of practice or 'constellations of practice' (Faulconbridge 2010), between which people (capable of abstracting ideas from situated contexts and carrying embodied skills), as well as documents (carrying information) and objects (capable of making knowledge architectures tangible, durable and contestable) can circulate (Cook and Brown 1999; Yakhlef 2010). 'Social validation' would then denote a process of incremental diffusion of ideas through multiple practice contexts. However, in this perspective another mechanism for the validation of knowledge would be underestimated: social institutions.

Institutions can be understood as socially shared formal or informal normative structures (rules), but also as cognitive structures, which limit choices and define roles as well as legitimate and appropriate action (Colyvas and Powell 2006; DiMaggio and Powell 1991). Institutions can provide career systems, which afford individuals limited mobility across contexts. In this way, hierarchical linkages between practices and competencies are created. Institutions can also provide procedures of validation. From a practice perspective, this means that specific practices are chosen to validate and legitimize the outcomes of other practices (Hutter and Stark 2015). For example, the clinical testing of pharmaceutical active ingredients is a precondition for their admission to the market. Thus, a specific and highly regulated routine at the intersection of clinical practice, experimental science and large-scale industrial project management is chosen to validate and legitimize the results of biomedical research.

Innovations can emerge from a multitude of social contexts. The question as to whether there are social institutions available to validate as well as economically exploit certain ideas is relevant. Müller and Ibert (2015) use this question to distinguish push- and pull-based innovations on the basis of the work of Hagel, Brown and Davison (2010). While in push innovations, ideas emerge from creative or scientific work and are converted into practical solutions via established mechanisms of validation and exploitation, pull innovations emerge in a more unguided manner. Examples for pull innovations are the creation of new legal advice services (Ibert, Müller and Stein 2014; Stein

2014) and the creation of new leisure or sports activities along with the appropriate equipment (Brinks and Ibert 2015). These innovations tend to be at odds with existing institutions, which is why they often fail to attract institutional support or funding. Rather than creating products for existing markets, they create their own markets – including the commercial players in them.

This differentiation between push and pull innovations, as well as the further differentiation into aesthetic (Reckwitz 2012) and rationalist modes of knowing as starting points for innovation (leading to a two-by-two matrix) was inspired by Amin's and Roberts's (2008) typology of knowing in practice. In it, Amin and Roberts suggest that innovations starting in creative and epistemic practices tend to be radical whereas application-driven innovations tend to be incremental. Müller's and Ibert's (2015) argument, which includes localized practices as well as social institutions, actually points in the opposite direction: Since push innovations unfold in a highly institutionalized environment, they might in fact be less radical or disruptive than innovations emerging from an environment which is not supposed to innovate.

In this framework, innovations in biotechnology are treated as push innovations – for several reasons: The key ideas are formulated as scientific discoveries and inventions. Intellectual property is protected by patents. Biotech companies as well as pharmaceutical and agro-industrial corporations convert the scientific ideas into marketable solutions. Legislation, regulatory authorities and, in the case of biopharmaceuticals, health insurance providers define procedures of validation and legitimization. The high degree of institutionalization of biotech innovation processes is also reflected by the fact that actors in the field use a sophisticated vocabulary, which delineates business models and development stages of new product developments (see for example Ernst and Young 2010). However, these concepts do not necessarily reflect the actual process dynamics of biotechnology innovation. Instead, they can be understood as a means of coping with an extreme degree of unpredictability and a high risk of failure for innovation processes. Within these processes, unforeseeable challenges, interruptions, dead ends, partial failures, accidental encounters, dire straits and dramatic reorientations are the norm rather than the exception, even if they turn out successful in the end (Ibert and Müller 2015; Mattsson 2009).

A practice perspective on knowledge-creation includes a novel way of thinking about space. Instead of territories or centre-periphery-gradients, the key spatial unit of observation is the site or locality of knowing in practice. Very roughly, this brand of spatial thinking can be labelled 'topological' or place-centred: Particular places provide material settings which support specific practices, for example 'the laboratory' (Livingstone 2003; Schatzki 2002). They are characterized by a typical presence, arrangement and interaction logic of human and non-human actors. In addition, they have a functional material order and operate under a typical time structure. Material places can be nodes in 'brain circulation' and thus are not only relevant for maintaining and performing practices, but also for reproducing communities and cultures (Hall and Appleyard 2009; Törnqvist 2004).

In order to accomplish innovation, multiple resources and knowledge practices across different places need to be combined, either simultaneously or sequentially. This is particularly true for biotechnology innovations, which require a large number of very different, highly specialized capabilities (Baraldi and Strömsten 2009). Strambach and Klement (2012) characterize innovation processes in terms of re-combinations of hitherto unconnected knowledge domains, often cultivated in distant places. They show that, as a cumulative effect over time, organizational routines and institutions change as new knowledge emerges from re-combination. Hence, innovation involves both boundary practice and re-negotiations of rules in social aggregates like organizations, institutional environments and communities.

Territorial innovation models such as the cluster or the regional systems of innovation approach have portrayed this interactivity as a spatially bound phenomenon (Moulaert and Sekia 2003). They highlight the importance of spatial proximity for collaboration. Relational economic geography has moved beyond this territorial fix, and developed ways to analyse relational and interactive dynamics in innovation processes crossing territorial borders and spanning distance. The 'buzz and pipelines' (Bathelt, Malmberg and Maskell 2004) literature, for example, traces the productive tension between specialized knowledge networks covering long physical distances on the one hand, and the thick, unpredictable and creative interactions in localized contexts in innovation processes. 'Buzz', understood as open and dynamic exchanges in ever changing constellations, is seen as the quintessential quality of creative places, which has been highlighted numerous times in accounts of creative or innovative milieus. By contrast, 'pipelines' refer to organized, selective, technologically aided transfer of knowledge over distance, as it is practiced in the corporate world.

According to Moodysson (2008), the importance of global pipelines outweighs that of local buzz in the case of biotechnology innovations. Especially the high degree of formal qualification as well as the exclusive and well-guarded membership roles in project teams, companies and research facilities, limit local buzz and at the same time support exchange over distance. While the biotechnology industry's strong propensity to cluster in very specific places is well documented, the reason for this needs to be sought in concepts other than 'buzz' (Howells 2012).

Aside from dyadic or multi-nodal ties over physical distance, places can be integrated into interactive dynamics of knowledge creation by a sense of belonging, or rather an ascription of belonging by knowledge practitioners. Crevoisier and Jeannerat (2009), for example, formulated the idea of 'multi location milieus' – knowledge milieus, which depend on place-bound interactions and highly specific contexts, but who find these conditions realized in a number of places and are mobile between them. In his study of knowledge practices in international architectural firms, Faulconbridge (2010) stresses that the mobility of architects – all but the highest-ranking ones – is in fact rather limited. Nevertheless, they use their preferred locations in internationally renowned metropolises as well as circulating objects and electronic communication, but also private travel, to create trans-locally shared notions of architectural performance. They do, therefore, represent one such multi-location milieu.

Similarly, Brinks and Ibert (2015) use the metaphor 'mushrooming' to explain the spatiality of entrepreneurialism originating in user communities. Through online communication, a latent web ('mycelium') exists between the countless localities distributed across several countries, in which users develop new practices. The creation, in specific places, of companies which manufacture equipment or provide services to the users then becomes co-constitutive to the emergence of a visible community. The development of such a community in turn brings about a new pattern of places, which house relevant activities (such as conventions and competitions).

Müller and Ibert (2015) present a generic conceptualization of spaces of practice by distinguishing localized 'communities of practice' and trans-local 'cultures of practice'. The former strongly reflect the ideas of situated practice formulated by Lave and Wenger (1991) and others. This level of analysis contributes to the study of innovation, in that it affords an observation of serendipitous events, encounters, experiences, irritation and inspiration between practitioners (of the same or of different communities) in specific localized, material settings. The latter reflect the idea of community as a social and cognitive realm of shared values and identity, of social belonging and common professional socialization. Examples are transnational communities (Djelic and Quack 2010) like specialized investors (Morgan and Kubo 2010) or political activists (Fetzer 2010), but also epistemic cultures like particle physicists and molecular biologists (Knorr Cetina 1999).

Trans-local communities or cultures of practice can be highly dynamic and transformative. According to Frickel and Gross (2005), SIMs resemble other social movements, as they are inherently political: They have at their core programmes for scientific and intellectual change, in particular 'intellectual practices that are contentious relative to normative expectations within a given scientific or intellectual domain' (p. 206). Hence, epistemic practices and orders themselves are subject to challenge as well as political and organizational mobilization. These changes also affect material orders, i.e. ordered relations of bodies. According to Vermeulen (2017), 'systems biology' is one such social and intellectual movement. A subtype of 'big biology' (Vermeulen 2010) its focus is on the shift from isolated, small-scale, hands-on biologic epistemologies ('reductionist biology') to an integrated, high-scale computerized modelling of cellular processes. In a process of circular reference, the movement lead to a spread of new forms of project organization and science policy models throughout Western countries, but also to the material creation of new sites of epistemic practice with fundamentally new internal, material orders.

This level of analysis focuses on socio-cognitive structures. Both levels are linked. It is best to view them in conjunction and with sensitivity for their interdependence. Again, Knorr Cetina's (1999) comparative study of the epistemic cultures of particle physicists and molecular biologists provides a template. Particle physicists, she argues, value standardized laboratory settings for their knowledge creation, identical across localities and free of any disturbance or deviation. They create large-scale research infrastructures like particle colliders and in turn, these are the spatial (topological) reference points in the global particle physics community. Molecular biologists on the other hand routinely create and exploit idiosyncratic settings in their laboratories, thus producing a more decentralized and diverse spatial topology.

Birch (2012) delivers an additional perspective on the key transnational dynamics in knowledge creation, specific for the case of biotechnology innovation: He argues that two forces are simultaneously at play: a) A localizing one, consisting of the flows and interactions crystallizing in particular places and generating idiosyncratic, partially tacit knowledge carried and practiced by a small number of people, and b) an abstracting one, consisting of the mechanisms of codification, regulation and standardization. The latter 'globalizes' the knowledge, but also enacts hierarchies and unevenness between places and territories, resulting from the dense interplay of corporations, lobbyists and regulators. Thus, power is introduced into the geography of knowledge creation.

Another conceptual advancement is the introduction of the time dimension into the discussion of localities. Territorial innovation models assume that permanent co-location is a prerequisite for the exploitation of physical proximity. One measure of physical proximity is the probability of a coincidental encounter between two people. This effect can be achieved by permanent co-location. Increasingly however, formats of temporary co-presence – conferences, trade fairs – are used to create opportunities for encounters (Schuldt and Bathelt 2009). Within these formats, increasingly sophisticated technical tools are used to optimize the chances for productive encounters between participants.

Furthermore, in the combination of the time and space dimensions, not only the advantages of temporary proximity, but also the advantages of temporary distance and spatial separation between localities become accessible. In their study of online collaboration in virtual hybrid communities, Grabher and Ibert (2014) show that the absence of co-presence frees participants from the need to adhere to a specified time-structure while contributing. In this and many other ways, physical distance or separation can be assets. During innovation processes, it frequently becomes necessary to split an innovative activity from an inhibiting environment, either because it does not welcome new ideas, or because the logics and practices performed in a place (e.g. a university) are incompatible with an innovative practice (e.g. running a start-up company) (Ibert, Müller and Stein 2014). For limited time-periods, innovation actors seek out localities, which are separated and only selectively connected to others. These can be laboratory situations with a high degree of experimental freedom, where aspects of the outside world are carefully 'imported' under controlled conditions (Latour 1999). They can also be highly specialized and standardized localities (like animal testing sites or clinical trials centres), where necessary validation and definition steps in an innovation process can be performed without interference.

Beyond distance and proximity in physical space, a number of scholars have created approaches and taxonomies to grasp and categorize other forms of proximity and distance, for example, institutional, cultural, organizational, technological, social, emotional and cognitive proximity and distance. All these concepts are ways to describe varying degrees of difference between participants in interactive, collaborative innovation processes, as well as their utility for innovation. According to Boschma (2005), institutional proximity is the umbrella term, while all other forms of non-spatial proximities are sub categories of the earlier. Knoben and Oerlemans (2006) by contrast arrive at the conclusion, that technological and organizational proximity and distance are the key concepts, which allow predicting whether participants in innovation processes can successfully collaborate or not. Among other things, they highlight the difficulties existing in collaborative relationships between large, more bureaucratic industrial organizations and small start-up companies.

Using proximity and distance as metaphors is a possibility for geographers to integrate conceptual developments from other disciplines concerned with the role of difference, most notably economic sociology. Sociologist David Stark (2009), for example, coined the notion of 'heterarchy' to denote functionally integrated social settings (a factory, a collaborative project), in which divergent orders occur in economic practice and need to be managed. These divergent orders are a source of both uncertainty and potential value. In the similar notion of 'structural folding' (Vedres and Stark 2010), individuals are simultaneously members of different internally cohesive groups (in the empirical example: game developers with different specializations) or 'multiple insiders'. Based on the different professional 'languages' used in different groups they create essentially new ones. Hence, difference can be a source of creativity.

One commonality across multiple approaches to difference in economic geography is that a distinction is made between spatial and non-spatial forms of proximity and distance. Furthermore, these approaches (in contrast to Stark) tend to favour proximity over distance. Proximity is seen as a prerequisite and nurturing factor for innovative collaborations, which are inhibited by an excess of distance. Newer approaches adopt a more differentiated stance. Trippl and Tödtling (2011) argue that innovation systems require a balance between proximity and distance. Like other authors, they present the optimal combination of proximity and distance in the form of an inverted U-shape. Along each dimension, an optimum is reached somewhere in the middle between zero distance and total separation. These approaches of course suffer from the difficulty to measure and quantify non-spatial distance, so the inverted U-shapes are to be understood as visualizations of an idea and not as exact diagrams depicting quantified relationships.

Ibert (2010) presents the notion of 'relational distance' as a heuristic concept to approach forms of difference in innovation networks in an open manner. Instead of providing a hierarchy or taxonomy of concepts, the approach includes all forms of 'cultural difference' as they are experienced and reported by persons involved in innovation processes. Using the concept as a heuristic tool, Ibert and Müller (2015) inductively identified six dimensions of proximity and distance which occur in innovation processes: institutional, organizational, cognitive, interest-based, functional, hierarchical and social (see Table 1.1). The authors do not claim comprehensiveness or complete absence of any kind of redundancy. Yet, in one particular way, this relational distance approach goes beyond the conceptualizations of proximity and distance discussed before. It adopts the idea of multiplexity (Haythornthwaite 2001), i.e. each relationship occurring in an innovation process can be proximate and distant along different dimensions. As a consequence, relationships can be considered productive in innovations, and can be assessed in terms of productivity, even if the intensities of proximity and distance occurring in them are much more extreme than the inverted U-shape model would suggest. Specific types of relationships can be named, characterized and studied for their specific contributions in different phases of innovation processes (see Figure 1.2). For example, the relationship of 'mentorship' is distant in terms of hierarchy (i.e. mentor and mentee differ in this respect), but proximate in terms of institutional and organizational context as well as

Table 1.1 Forms of proximity and distance in innovation networks

Dimension	Proximity	Distance
Cognitive	Same disciplinary enculturation and cognitive patterns	Different disciplinary enculturations and cognitive approaches
Organizational	Same organizational/sub-organizational affiliation	Interaction crosses organizational/sub-organizational boundaries
Institutional	Same or similar rules, norms and conventions	Different or dissimilar rules, norms and/or conventions
Social	Personal acquaintance and commonality beyond the professional sphere	Only superficial acquaintance and no or little commonality beyond the professional sphere
Functional	Mutual accessibility due to interrelated roles, routines or assignments (e.g. provider-customer relation)	Unrelated roles, routines and assignments, no mutual accessibility in everyday routines
Interest	Shared interest in an idea, willingness to share risk	No shared interests, risk-averseness
Hierarchical	Comparable access to organizational resources, occupying positions on same hierarchy level	Different levels of access to organizational resources, occupying positions in respective hierarchy

Source: Ibert and Müller 2015

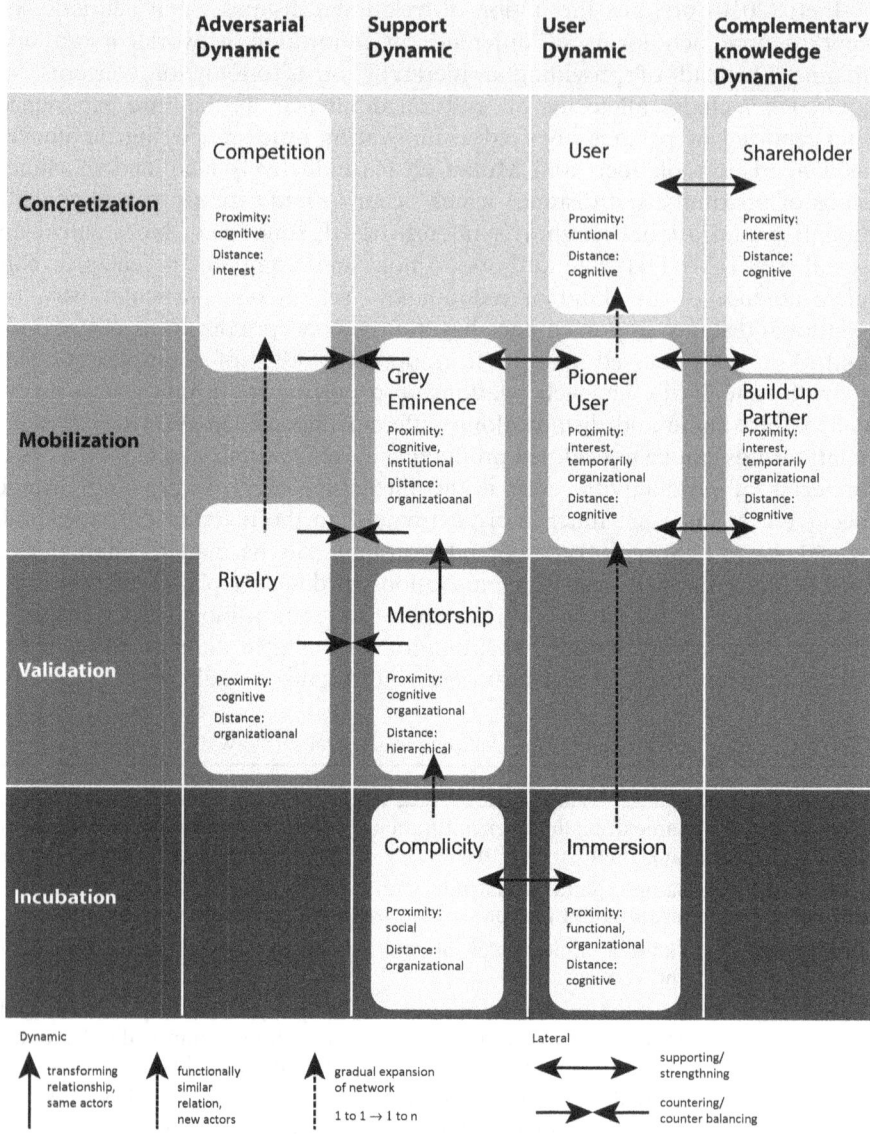

Figure 1.2 Relationship dynamics in innovation processes
Source: Ibert and Müller 2015

knowledge domain (cognitive). This relationship enables innovators to validate their ideas experimentally in a safe environment (Ibert and Müller 2015).

Furthermore, each relationship's properties of relational proximity or distance manifest themselves in physical space. Physical distance is, thus, not treated as a separate category. Instead, relational proximity and distance are understood to play out in activity patterns in time and space; therefore, physical space is seen as the enabling substrate of relational distance or proximity. As shown above, physical distance can have positive effects on innovation. Relational distance, too, has a function in innovation networks. As a property of multiplex relationships, even a very high degree of relational distance can become productive, if it is complemented by proximity in other respects (Menzel 2015).

This is not an argument along the lines of physical proximity compensating cognitive distance. Instead, every relationship contributes a unique transformative element to the overall innovation process, which is rooted in its specific combination of proximity and distance. For example, the 'pioneer-user' relationship has to be highly distant in cognitive terms; otherwise it could not contribute the essential transformation from an idea which works in an experimental environment to a solution which works in a context of application. What holds this relationship together is the mutual interest in the realization of an idea (Ibert and Müller 2015).

Some relationships are overall more distant than others. Another contribution of the relational distance approach is that it appreciates the value of such mostly distant or even conflict-ridden (sometimes hostile) relationships. For example, the relationship of 'rivalry' helps innovators to refine their idea, to make it robust and to formulate negative demarcations. It enables them to state in what way they want to do things differently and from whom they want to set themselves apart. In the course of an innovation process, relationships gain and lose importance. Some are highly temporary in nature; others persist but fade into the background and change in their meaning (for example 'mentoring' and 'grey eminence' relationships). In other cases, a relationship is followed by functionally equivalent relationships, but with different participants (for example 'pioneer-user' and 'customer' relationships).

1.4 The relational work of translating in innovations

A field as laden with tension as biotechnology – highly bureaucratized and yet heavily resting on individual entrepreneurship, science-driven and yet extremely speculative, mobile and yet materialized in an extremely uneven geography – provides an opportunity: An opportunity to bring the agency of relational work into economic geographies of innovation processes. How are relational distance and proximity enacted? How are relationships built, maintained and ended? What is the effect of selective choices – relationships entered into vis-á-vis relationships not entered into – both for an innovation and its relational environment? The elaborations above made clear that categories such as relational proximity and distance, applied empirically, are somewhat technical, functionalistic and not free of contradictions. They rest on exaggerated assumptions of objectivity and structural determination (two actors *are* different, depending on their positioning in institutions, cultures or organizations). In this section the perspective will be expanded in a three-step kind of way. First, I will discuss sources of difference. Then, I will discuss the relational work of both building and ending relationships. Finally, I will discuss the two-way impact of such endeavours: on an emerging innovation, as well as its environment. The aim is a conceptual basis for future taxonomies or typologies which do not distinguish types of relationships, but rather types and logics of relational work in innovations. Investment, discussed in the following section, shall be accommodated in this framework.

There is general agreement that innovation and science-driven entrepreneurship (as evident in biotechnology) involve agency across institutional divides. Powell and Sandholtz (2012) refer to biotechnology entrepreneurs as 'institutional hybrids' or 'amphibians'. According to Boschma (2005), institutional difference is the fundamental form of relational difference, with language being a key carrier and manifestation of institutions. Sociologically defined, institutions represent the cognitive orders shared in a society: They are ascriptions of appropriate and legitimate actions to types of actors. In this approach, actors adhere to role-definitions which are acquired through socialization and cultivated in shared expectations. A large part of this role-knowledge is 'sedimented' or 'taken for granted' as far as they concern human behaviour, as well as the normative (Colyvas and Powell 2006; DiMaggio and Powell 1991; see also Bathelt and Glückler 2013). However, this perspective is somewhat biased towards shared normative structures.

The focus here is more on socially shared orders of valuing actions, practices and objects. According to David Stark (2009) the concept of 'orders of worth' is a generic approach to orders of economic value creation. Its defining feature is that it conveys the inseparability of the two meanings of value: values are normative constructs, but value also refers to the potential gain an object or action yields. The two aspects always occur in conjunction, as every gain of something valuable is also a moral gain (e.g. being a successful entrepreneur), and thus invokes normative values. Normative values, in turn, are legitimized and reproduced through their practical utility (e.g. property rights produce wealth), hence their instrumental value. Orders of worth are thus socially shared patterns of valuation and justification which link the moral and instrumental value of actions, practices and objects (my definition, based on Stark 2009 as well as Boltanski and Thévenot 2006). Boltanski and Thévenot 2006 identify seven 'worlds', at least three of which are likely to be highly important for biotechnology innovation: The 'industrial world' in which rationality, efficiency and formalized organization are key values; the 'world of fame' in which individual fame is the key value; and the 'civic world', in which societal responsibility is the key value. In biotechnology these orders might manifest themselves in the idea of community (value: sharing of expertise and resources, especially in epistemic communities, potentially also user communities), the public research system (value: public knowledge) and the regulatory framework (value: safety, bureaucratic correctness).

Innovation entails *knowledge creation*, but also *knowledge evaluation*. Adding to confusion it involves *evaluating and appreciating ways of creating knowledge*. Appreciating knowledge-creation entails preferences for questions and objects of knowing, preferences for applicability of knowledge in specific practices and preferences for particular methods of knowledge-creation. It also includes different forms of non-knowledge: recognition of the boundaries of what can be known and the limitations of one's own practice, but also delineations of knowledge practices and objects which are deemed useless or counterproductive (Knorr Cetina 1999). Along with the appreciation for knowledge, actors have different conceptions of value and different practices of assessing it. Their application in action situations depends on their adequacy as perceived by actors. Sometimes actors are unclear as to which rules and associated practices of knowledge creation and value assessment currently apply. According to Stark (2009), these are situations of uncertainty. Thus, uncertainty is not defined by a lack of specific knowledge ('not enough data'), but by absence of a clear order. Despite its intuitively negative connotation, the ability to experience and exploit uncertainty is considered a potentially benefiting factor for innovation, as it allows the introduction of new orders of worth to value creation. In situated contexts, difference can be experienced. The coexistence of divergent orders in one local context is denominated as 'heterarchy' by Stark (2009). This can be the case, for instance, when the hierarchical order of an organization collides with the peer-oriented order of an epistemic community in a corporate R&D department.

How can individuals relate to the orders in which they operate? Institutional approaches are criticized for understanding individuals as 'institutional dopes' (Lawrence, Suddaby and Leca 2010). More recent approaches, therefore, see the relationship between practice and institution as a more reflexive one. In the 'institutional logics' approach, a reflexive layer is put between practice and institutions, allowing individuals to interpret and change the relationship between them (Thornton, Ocasio and Lounsbury 2012). One approach that deals specifically with conscious, radical and successful institutional changes is that of 'institutional entrepreneurialism' (Battilana, Leca and Boxenbaum 2009; Crouch 2005). Institutional entrepreneurs recombine institutional orders in a previously not conceived manner. More precisely, they apply a particular institutional order (for example industrial certification) along with its specific governance forms and practices to a societal task, which hitherto has been within the realm of another order (for example public colleges; Crouch 2005; Thornton, Ocasio and Lounsbury 2012). Institutions hence do not automatically translate into specific practices. Instead, drawn-out processes of situated interpretation and application can lead to highly diversified and unpredictable results in terms of practice. Moreover, orders are not monolithic blocks, but in themselves relational. They are cultivated in social collectives (nation states, communities, organizations etc.), within which and between which actors can position themselves.

One classical notion of positioning between worlds is that of 'brokerage' originating in network research (Burt 2004; Obstfeld 2005): Individuals with connections into different social groups can benefit and create value by connecting the two; either by exploiting the difference as a form of arbitrage ('tertius gaudens') or by building new lasting ties ('tertius iungens') (Klagge and Peter 2009). Vedres and Stark (2010), using the video games industry as an example, go one step further by focalizing 'multiple insiders' in 'structural folds'. These actors are not in-between, but rather fully integrated in several (at least two) culturally coherent groups, e.g. communities of practice. These individuals are prone to contribute to innovation and creativity, as they translate between cultures and languages and thereby create something new. The authors stress the creative character of this type of relational work: The outcome is in no way predetermined, but a highly individual and unpredictable creative product. This finding indicates that the productivity of relational distance is in the eye of the beholder. Relational distances are enacted and constructed in a situated way, depending on subjective perception. Some historical examples illustrate the importance of situated interpretation.

In their account of the Medici family's rule in Renaissance Florence, Padgett and Ansell (1993) conclude that actors achieve lasting influence and power in a locality by strategically combining participation in various 'games' (ordered fields of interaction), some typical for aristocrats, others more common. Specifically, the Medici managed to act in such way that every 'move' made sense in more than one game. The authors highlight the localized entanglement with multiple orders and the local situatedness of action, which they termed 'robust action'. Another example stems from the history of financial products. In the 19th century life insurances were introduced to the American market. The first attempt to sell such financial products failed, however. The deeply religious and conservative populace rejected the instrument, because it was perceived as a form of gambling on life and meddling with god's plan. Several decades later, life insurances were successfully launched by appealing to these exact values. Now the ideal of the prudent, hardworking (faithful) family man was invoked, who would take all measures necessary to secure his family's wellbeing (Zelizer 1978).

In addition to understanding how individuals position themselves in (or between) orders, interpret and align them, an approach is required to conceptualize the relational work of building and changing relationships. Existent approaches in economic geography and sociology effectively conceptualize processes of search. In Ibert and Müller's (2015) phase model, in innovation's 'incubation phase' serves to create meaningful (yet often unrecognized) cognitive dissonances which lead to the creation of new ideas. Brinks and Ibert (2015) speak of 'tinkering' when describing the productive kinaesthetic interaction between humans and objects at the beginning of a user-driven innovation process. Rutten (2017) uses the concept of 'conversations' to describe how a problem is gradually defined and framed through communicative interactions across places and communities. Heuristically a distinction between 'local search' and 'non-local search' can be applied (Rosenkopf and Nerkar 2001). This distinction can be understood as being both relational and material: 'local search' could mean that the time-spatial activity pattern needs to be changed only to a small degree, while in the case of 'non-local search' it is changed substantially. What these approaches miss, however, is the degree to which the creation of relationships to pursue an innovative idea (once clearly defined) involves struggles: Each established relation replaces an existing or a potential one. This is particularly true for later phases (validation, mobilization, concretization – Ibert and Müller 2015). Investors in particular choose selectively, allowing one innovation to flourish while preventing countless others.

Stemming from STS, the theory of 'translations' formulated by Michel Callon (2007) is a very practical way to trace such processes. In contrast to its intuitive meaning, a translation in this theory is not just a transfer of meaning into a different context. Instead, in the course of a translation, an actor achieves an alignment of other actors under a purpose provided by him or her. He or she manages to replace their multitude of voices and assumes the role of a speaker for the whole. A translation is thus a replacement of expressions and an alignment of actors under a new collective construction of purpose and truth (ibid.). According to Callon, a translation process has several 'moments', i.e. instances in which the process unfolds. These moments can overlap and are difficult to discern as temporal phases. Callon uses a case study of a group of scientists who attempt to implement a Japanese breeding method for scallops in the Bay of Saint-Brieuc, France, to demonstrate the translation process.

The opening act of a translation is called problematization: an initiating actor (in this case a group of scientists) proposes that other actors are connected to one another by a joint problem as well as a potential solution. In this act, the other actors as well as their relations receive a definition, an identity, a role. The initiators make themselves inevitable by defining 'obligatory passage points', key questions which only they can address, and which the other actors are unable to avoid due to 'obstacle problems', likewise defined by the initiators. A problematization thus contains a novel problem definition (an 'innovative idea' in the terminology of Ibert, Müller and Stein 2014), but it is more than that. It also contains identity and role definitions as well as a proposed order of relations between actors. In this particular case, the scientists tried to convince fishermen (and scallops) that they would soon face the extinction of the local scallop population if they continued with the current catching technique. A new breeding technique, which the scientists had imported from Japan, would remedy this situation.

All entities involved can refuse this proposition through action. In the ensuing 'series of trials of strength' interessement is the 'group of actions by which an entity [...] attempts to impose and stabilize the other actors it defines through its problematization' (Callon 2007, p. 62). The term interessement is derived from Latin 'inter esse' – 'to be between'. In a quite literal understanding, 'to interest other actors is to build devices which can be placed between them and all other entities who want to define their identities otherwise' (ibid., p. 63). Hence, it describes both strategies and technologies of attracting and convincing actors. The role definitions proposed by one actor compete with other definitions proposed by other actors. The notion of interessement thus recognizes the limited number and mutual exclusion of possible translations. It is about separating actors from competing problematizations as much as it is about association. By employing the term 'interessement device', both material objects and practices or performances as elements of relational work are addressed. There is a certain similarity to the concept of 'boundary object' or 'boundary practice' respectively. However, while the latter are used to explain stable, routinized interactions across sociocultural boundaries, interessement refers to activities of purposeful change. Strategies and practices of interessement, understood as a form of relational work, will undoubtedly have specific geographies. They will involve travelling artefacts and humans, patterns of co-presence and co-location, as well as dependence on particular places ('buzz') and connections ('pipelines').

Successful interessement achieves 'enrolment'. The term refers to the 'negotiations, trials of strength and tricks' (ibid., p. 66) which enable interessement efforts to succeed, as well as their result: stable relations expressed in statements, which replace questions and uncertainties. This stability lasts until challenged by a new translation effort. Being a representative of Actor-Network Theory (ANT), Callon maintains that there is an ontological symmetry between human and non-human actors (in this example fishermen and scallops); both require enrolment. This ontological symmetry provides a possibility to conceptualize two key challenges of innovation on one level (Akrich et al. 2002): Like Callon's scientists, biotech innovators need to engage with human actors (investors, customers) and non-human actors (cells, genes). Both are highly unpredictable, both need to be stabilized in roles, identities and relations, which cannot be taken for granted in advance. As in the case of interessement, a situation of enrolment will be kept stable by spatial arrangements such as trans-local connections and mobility patterns.

This way of introducing ANT thinking into relational economic geography is not completely new, and it serves a recognized purpose. The field of ANT, which in fact is a roof category for a broad and diverse range of materialist and post-structuralist approaches, has had influence in human geography for some time (Müller 2015a; Murdoch 2006). 'Flat' or post-structuralist geographies are the subject of lively debates, one of the key issues within which is the conceptualization of space. Post-structuralist geographies operate with topological and relational ideas of space rather than scalar and territorial notions (Marston, Jones and Woodward 2005; Jessop, Brenner and Jones 2008).

The first purpose in this context is to describe emergent innovations as actor-networks. While Ibert and Müller (2015) characterize emergent innovations ad 'idea-centric networks' focusing on relationships between people across organizational and institutional boundaries, I wish to include the, as it were, content, essence or materialization of the innovation as well. ANT in part emerged from the field of STS as well as Laboratory Studies (Knorr Cetina 1984; Latour and Woolgar 1979), which study the situated, material conditions of knowledge-production in science. This origin represents, as it were, a natural linkage to the study of innovation processes. ANT-based thinking can help to understand the emergence of technological structures in a relational perspective. Traditionally, technology is a difficult subject for economists and social scientists. According to Dosi and Grazzi (2010), technology consists of procedural rules, knowledge, artefacts and physical inputs (like energy). These elements are linked to achieve a particular purpose defined by humans. Similarly, Arthur (2009) defines technologies as hierarchically integrated systems, which harness naturally occurring phenomena to serve a specific purpose. These conceptualizations are still ontologically hierarchical and not 'flat'.

One of the most conspicuous features of ANT is the way action and actors are theorized. While more traditional approaches ascribe the ability to act either to individuals (humans) or to collective actors such as organizations, ANT views action as a process which is distributed across human and non-human actors (Latour 1996, 2005). The action 'to shoot', for example, is distributed across a human body and a gun. Both together form an actor-network. The fact that guns are seen as (partial) actors and not just as 'tools' illustrates the potential for controversy which this approach brings (Latour 1994). An emergent innovation can be understood as a growing and changing actor-network composed of human and non-human entities as well as practices (professionals in a lab, dealing with specimen, wrestling with the unpredictable behaviour of their epistemic objects, creating data, constructing new artefacts etc.). Such an actor-network constitutes the materialization of an idea and the material substrate for its refinement. In fact, the idea only exists in this actor-network.

The second purpose is to describe the relations between the innovation itself and its environment: What change does an innovation effect to companies, markets and user communities? Popular structuralist conceptualizations of change, such as 'structuration' and 'hysteresis' (Barley and Tolbert 1997), stress gradual change through repetitive routine. In a flat, post-structuralist ontology, change through action is the norm rather than the exception. ANT focuses on associations: activities during which individual actors, human and non-human, are materially connected and their connections are stabilized (Latour 2005). Such associations can create socio-structural entities (organizations, communities, nation-states), whose survival requires constant work and is constantly challenged by competing associations. Large corporations, for instance, consist of the interplay of myriad human and non-human entities – communication networks, control mechanisms, buildings, images, text etc. As a consequence of this, Callon and Latour (1981) perceive the 'big leviathans', as they call large structural units, as rather more unpredictable than they would be if there were such a thing as a genuine 'macro-level'. Martin Müller, too, urges us to open up the black box of organizations (Müller 2012) and not to presuppose their role as actors. Thus, the foundations of the power of organizations, particularly large transnational corporations, becomes palpable instead of being assumed as often happens in critical geopolitics (ibid.).

Markets, likewise, are not seen as abstract entities in the neoclassical sense ('the meeting of supply and demand'), but as products of material associations. They can grow out of communities of practice, i.e. from non-market forms of economic value creation, but they require action to be created. They require the coordinating work of humans together with measuring and calculation systems, communications networks, technical interfaces, places of structured face-to-face interaction, logistics systems etc. Hence, markets and market actors are materially constructed (MacKenzie 2009). The working of these hybrid assemblages which create markets is termed 'agencement' (Callon 2008). Markets are subject to constant observation and analysis. Observation and analysis devices are not independent or separate from markets but rather part of the agencements which constitute markets. The effects of such devices are therefore considered 'performative': they make market reality by analysing and representing it (Callon 2007; Callon and Muniesa 2005; MacKenzie 2008).

These approaches are highly relevant for the study of investment in biotechnology. They show how closely the production of markets and the production of knowledge are related (Glennerster, Kremer and Williams 2006; Walsh 2002). They also show to what extent valuations and actions outside the laboratory context feed back into science and participate in knowledge-creation (Hoof, Jung and Salaschek 2011). Haller (2011), for example, shows how cortisone as a therapeutic agent emerged from complex and unpredictable interactions in industry, clinic and science.

The terms agencement or 'assemblage' and actor-network are often used more or less as synonyms (Müller 2015a, 2015b). Here, I will use the term actor-network to denote associations of material bodies, which act together, i.e. across which action is distributed. Sometimes the concept will be used together with the attribute 'hybrid' to stress the diversity of involved bodies (human, non-human, living, 'dead'). The term assemblage is used to denote a wider set of dynamically unfolding associations, which can involve entities of a very different ontological nature (e.g. objects and ideas) and enact wider social realities such as markets and technological paradigms. According to Müller (2015b), 'assemblage' is a translation of the French 'agencement'. It refers to processes of socially and spatially distributed production of social realities (e.g. markets). Hence, during innovation, new forms of economic value creation are produced by establishing new products on existing markets, altering markets or creating new markets and associated social practices. In terms of geography, each process of assemblage will involve its own set of place-bound interactions and connections across places. Furthermore, Müller (2015b) argues that every dynamic of assemblage has the potential to attack existing forms of territoriality and at the same creates new forms of territoriality. An innovation process can be described both as an unfolding actor-network (the interconnected actors materializing the emergent innovation, making it 'happen') and in radical cases as an emergent assemblage (a new social reality like a new market segment or a fundamentally new type of medical therapy).

Thus the three-step is complete. To sum up: How do (human actors) position themselves in relational constellations and how do they interpret, (re-)align and apply the orders of worth under which they operate? How do they engage in search and later interessement/enrolment, building selective relationships and actor-network? And finally, how do the results of these activities spread in terms of larger assemblages? Rather than describing all actors and interactions involved in these dynamics individually – which would obviously be impossible – I seek to identify typical, recurring dynamics of relational work, actors who perform them and the logics they adhere to.

1.5 Integrating investment relations

The last section introduced dynamics of relational work and unfolding relational webs materializing innovations. In this context, the aspect of finance and investment is not appropriately reflected. While the relationship between business life cycles and venture capital is well understood, the role of investment is less clear if the firm level is not taken as the only reference point. How can investment in a 'financialized' sector be studied, if the usual levels of analysis are absent? In this study, investment is studied in the widest possible sense as a form of relational work, carried out by more or less specialized actors, during which financial (possibly along with other resources) are allocated to advancing an innovation in order to create future financial (and possibly other) gains based on *some* metric of value through which potential gains are weighed against a perceived risk (the latter already representing a form of socio-cognitive ordering).

Indeed, the financing model which is most strongly associated with supplying innovation-oriented financing in general, and biotechnology in particular, is venture capital. Since VC has been studied extensively, the role of VC investors will in all likelihood not yield many surprises: It can be expected to be reinforcing of existent structures and relations. Yet, on its fringes, the VC model has openings.

Kenney defines venture capital investors as follows: '[…] In a nutshell they are financiers who invest equity capital in young firms in the hopes of receiving an out-size return provided the firm rises dramatically in value' (Kenney 2012, p. 61). Venture capital investors are financial intermediaries in a twofold sense (Klagge and Peter 2009). They mediate between entrepreneurs who require start-up or growth capital and institutional investors such as banks, insurance companies and pension funds, but also corporations and wealthy individuals. They do so by setting up and managing closed investment funds with durations of typically no more than ten years. VC is considered a high-risk high-gain asset class. Companies which receive venture financing typically would not be funded by any other type of financier (e.g. banks). Institutional investors who invest in venture capital expect a return of 10 to 20 per cent p.a., but also face the risk of substantial losses.

Venture capitalists invest funds in innovative, partially newly founded ('early-stage') companies by acquiring equity in them. They try to minimize the risks associated with their investment activity by a) creating a portfolio and thus distributing risk across multiple investments, b) selecting investment objects carefully and c) taking a very active role in company strategy and management, thus guiding investee firms towards a profitable exit (Hellmann and Puri 2002; Lehtonen and Lathi 2009). VC is allocated in 'rounds', which are typically denominated alphabetically. In an ideal typical sequence, a start-up company with little more than a team and a technological idea receives a small A-round to produce a working prototype or another form of proof of concept. Further validation and adaptation to market requirements is funded with a larger B-round. The C-round serves the financing of a market launch, which is often vastly expensive.

An exit, i.e. the disinvestment of a venture capital investor, can take various forms. A company can be sold to other venture capitalists or private equity companies, which have a financial interest. It can also be sold to a corporation which has a technological interest (trade sale). Rather than selling to a specific buyer, a company can be made public, after which shares are sold profitably. Finally, a company can also pay off the investment gradually based on actual profits. This dynamic reflects another aspect of venture capitalists' role as intermediaries: Venture capitalists valuate companies or individual innovations in companies, which are not yet established on the market, based on their anticipated market value. This anticipation of future market value and the accompanying selection and steering activities in investment relationships are seen as the unique quality and competence of venture capitalists (Berglund, Hellström and Sjölander 2007).

It is noteworthy that the innovative effect of venture capitalism is most clearly visible if it is analysed on an ecologic level – that is, focusing on venture capitalists as elements of regional networks or in the context of nation states. The particular nation state in question is of course the United States, where venture capitalism was both created and cultivated as an engine of disruptive innovation. The region which is most commonly cited when the impact of venture capitalism on new technologies is discussed is the San Francisco Bay Area or 'Silicon Valley'. According to Zook (2004) as well as Ferrary and Granovetter (2009) the persistent performance of Silicon Valley in creating innovative technologies is partially due to the way venture capitalists are embedded in regional networks. Specifically, venture capitalists in Silicon Valley cultivate dense and diverse relational ties with entrepreneurs, other investors as well as universities and more mature companies. These put them in the position to select the most innovative start-ups and aid their development towards the market not only with business skills, but also with a profound understanding of technologies and with high-profile relationships.

However, in no other region could a comparable density and dynamism be found (see for example Pinch and Sunley 2009). According to Hochberg, Ljungqvist and Lu (2007) the quality of network relations, in which a VC is embedded, are critical to investment success. But this embeddedness is not necessarily the consequence of conscious action. Thus, from a relational and ecological perspective it becomes difficult to assign VC the role of a cause or independent variable for innovative success. Instead, the participation of venture capitalists in successful, disruptive innovation could be the consequence of other influences, such as the central positions of some regions in an international financial regime (Zeller 2003).

Several studies indicate that the role of venture capitalists is more ambiguous when viewed on a micro level. One claim in particular might be difficult to defend: venture capitalists' ability and inclination to identify and engage with radically new ideas. According to Miloud, Aspelund and Cabrol (2012) venture capitalists focus strongly focus on a number of criteria when evaluating investment candidates: attractiveness of the respective industry, quality of founders and management and quality of external relationships rank highest. One environment where ideas and entrepreneurs are selected by investors are investor conferences and 'pitch' presentations. Clark (2008) found that venture capitalists are strongly influenced by the oral presentation performances and presentation skills of entrepreneurs, but are unaware of it and instead attribute their choices to facts.

According to Lehtonen and Lathi (2009) entrepreneurs can drastically increase their chances of acquiring VC if they employ advisors to optimize their proposals. While this finding is not surprising, it points at an element of standardization in investment selection. Stuck and Steingarten (2005), although being venture capitalists themselves, are highly critical about the innovative performance of their trade. According to them venture capitalists are not particularly apt at talking to technology-driven inventors and rather cultivate a business habitus. They find that VC is ineffective in financing truly innovative ideas for several reasons; the life cycle of a venture fund does not allow investment durations substantially longer than five years due to the time-consuming nature of both pre-investment activities (fundraising, selection of investments) and post-investment activities (exit, pay-out) in the ten-year timeframe. This time span does not allow lengthy research and development activities. Furthermore, venture capitalists tend to orient their activities both to market cycles and peer behaviour. Finally, in order to understand and assess an entrepreneur's business case, comparable business models and technologies must exist as a reference.

In addition to investment selection, the way in which an investee company is developed subsequently is relevant for the innovative outcome. Kortum and Lerner (2000) found that, while VC funding is only weakly associated with an increase in R&D expenditure, it leads to a substantial increase in patenting. Thus, an emphasis is put on exploiting existent knowledge. There is substantial controversy regarding the question as to the way in which venture capitalists add value to an enterprise (Large and Muegge 2008). Berglund, Hellström and Sjölander (2007) point out that next to productive learning processes induced by venture capitalists (based both on analytical reflexivity and practical problem solving), there is also a transfer of unreflected assumptions in investment relationships. Clearly, VC investment leads to a professionalization in investee companies. According to Hellmann and Puri (2002) this refers most frequently to human resource policy, stock option plans as well as the hiring of marketing professionals and professional CEOs to replace founders. In sum, in the terms of neo-institutionalist theory (DiMaggio and Powell 1991), VC investors can be understood as part of the institutional environments of young companies, and they exert legitimization pressures on them, guiding them towards a 'normal' organizational form.

These norming processes can create conflicts, which typically are perceived as stressful and unproductive by entrepreneurs (Zacharakis 2010). According to Ţurcan (2008) 'goal alignment' between venture capitalists and entrepreneurs produces highly differentiated results in the perceptions of entrepreneurs, from 'life changing opportunity' to 'enslavement'. Cumming found that the more power venture capitalists exert on an investee company (measured in contractual rights), the more likely the occurrence of an exit in which the company is sold to a corporation instead of remaining independent and either paying off the investment or going public (Cumming 2008). This indicates that venture capitalists, given the choice, would rather follow the safer path of selling to a corporation (which can be done in an earlier development state) than taking the risk associated with building an independent, publicly listed company.

In economic geography, the most intensively discussed question regarding VC is the possibility of an 'equity gap', that is, a supply gap in investment capital between more central and more peripheral regions (Chen et al. 2010; Klagge and Martin 2005; Martin et al. 2005). At the core of this discussion is the assertion that VC investment is highly distance-sensitive. VC investors are intimately involved in the management of investee firms. While this does not necessarily entail the steering of day-to-day operations, it includes activities such as participation in board meetings and informal discussions with entrepreneurs. Venture capitalists have a professional desire to stay in touch and be informed about recent developments in their portfolio companies, a need that requires frequent co-presence, sometimes at very short notice. This practical necessity has led to the proliferation of various heuristic notions about the distance which can be spanned in a VC investment relationship. The most popular is the 'one-hour distance rule', which states that a venture capitalist will only invest in a company which can be reached within one hour (Griffith, Yam and Subramaniam 2007).

Empirically, this rule was not found to be an absolute. Nevertheless, physical distance is a relevant factor for VC investments. Focusing on venture deals in Silicon Valley Griffith, Yam and Subramaniam (2007) could show that, depending on the dataset used, 40 to 60 per cent of all deals of regional VC companies are closed with companies either completely localized in the region or with their headquarters there. Specifically focusing on biotechnology companies, Kolympiris, Kalaitzandonakes and Miller (2011) find that there is a strong relationship between VC investment and firm location. This relationship, measured in spatial autocorrelations, was significant and positive for a company's spatial distance to VC companies and to other successfully financed biotechnology companies – but only within a ten-mile radius. This notion of a strong association of close spatial proximity and VC investment is very much in line with the prevalent cluster-oriented thinking with respect to biotechnology in economic geography.

In discussions of differentiated VC dynamics in central and peripheral regions ('equity gap'), two core lines of argument can be discerned: a supply-oriented and a demand-oriented one. The supply-oriented argument focuses on the availability of VC investors in regions. Several studies found that VC companies are highly concentrated in regional and national financial centres (Martin, Sunley and Turner 2002). In its simplest form, the question which is asked is: can such centrally located venture capitalists span the spatial distance to investee firms in more distant locations? This question is typically answered with a 'yes'.

Fritsch and Schilder (2006) argue that the key mechanism to explain the spatial reach of venture capital is 'syndication' (Lerner 1994). In most cases, venture capitalists do not enter an investment alone, but instead form a consortium of several investment firms. Peripheral regions lack high concentrations of VC companies and thus the financial power for large-scale investments. However, smaller VC offices and more mentoring-oriented private investors ('business angels') can usually be identified, who are adapted to the practical challenges of investing in the periphery. Syndication allows venture capitalists to combine these local contacts with the bigger financial scale and other resources of more centrally located investors. In such syndicated constellations, a regional or local venture capitalist can act as a lead investor and realize all day-to-day interactions with the portfolio company. According to Fritsch and Schilder (2006), syndication as a mechanism is sufficient to span spatial distances and hence – with regard to Germany – there is no equity gap (see also Bannier and Grote 2008). Syndication is not only a strategy to span distances. It also allows investors to combine non-financial resources such as specialized knowledge and relationships. Consequently, syndicated investments are on average more successful than individual investments. Furthermore, syndication allows investors to participate in more investments with smaller individual investment sums, which allows for a better distribution of risk within investment portfolios.

Most studies on the location of VC companies and the dynamics of VC investments arrive at the conclusion that venture capitalists follow attractive investment opportunities. Accordingly, the availability of 'investor ready' companies is decisive for VC investment and not vice versa (Rosiello and Parris 2009). Rosiello and Parris focus specifically on health-oriented biotechnology in the UK. According to them, the technological dynamism in regions drives VC investment. Consequently, concentrations of VC companies emerge in areas with dynamic industries and technologic opportunities. According to Chen et al. (2010) VC companies which are located in VC centres outperform their counterparts in more peripheral locations. Interestingly, centrally located venture capital companies can also have physically distant investments and still outperform their peers in the periphery.

While Klagge and Martin (2005), essentially employing a more supply-oriented argument, argue that public institutions such as public regional venture capital funds and other forms of financial regulation can effectively counter the equity gap, the demand-side argument leads to quite different policy recommendations. Here, the fostering of innovative industries and technologies combined with incentives for distant investors to engage with them represent the most sensible path of investment policy for peripheral regions (Chen et al. 2010).

However, the finding of a financial centre bias does not only apply to the supply side. Being located in a financial centre also has a noticeable influence on young companies' chances to be successful. Specifically, the chance of successfully going public is substantially higher for companies located in financial centres (Wójcik 2009). Explanations for this bias are the presence of specialized intermediaries and the existence of highly concentrated specialized labour markets in financial centres. As a consequence, young firms are more likely to count as successful investments and thus attract further VC into the region if they are located centrally. According to Klagge and Martin (2005) the existence of local stock exchanges is an important factor, which underpins a regional VC industry, not because portfolio firms are necessarily traded there, but because the local presence makes it easier to engage with financial practices and build relationships into the financial sphere.

These findings warrant a reconsideration of the duality of supply- and demand-side arguments and specifically the dichotomous manner in which the two sides are treated. Therefore, I perceive two contributions as particularly inspiring for this study, in which spatial distance is contrasted with relational proximity (Ibert 2010) and in which the negotiation and opening up of social orders through relational work is highlighted. Saxenian and Sabel (2008) analyse the role of first-generation Taiwanese immigrants working in US high-technology industries in the creation of VC dynamics in their home country. Being enculturated in both national environments, they have the capacity for a more reflected and searching approach to the problems of institutional development. Using their relationships, they participate peripherally in the identification and negotiation of potentials for the creation of VC institutions. This study indicates that practice can be shared over long distances, and that the cultivation of relational proximity over physical distance offers new perspectives for engaging one's physically local environment.

Wray's (2012) account of the VC communities in the North East of England and the East Midlands points in a very similar direction. In comparing the boundaries and outside relations of both communities, she comes to the conclusion that the cultivation of relationally proximate ties over physical distance (to other venture capitalists) also positively affects VCs' ability to engage meaningfully on a local level. She found that the VC community in the North East of England is comparatively closed and cultivates few outside ties, while venture capitalists in the East Midlands maintain multiple ties to colleagues in other regions. These external ties are a resource which is used to foster activities in their own region – a dynamic which is largely missing in the North East. Accordingly, venture capitalists in the East Midlands engage with (actual and potential) entrepreneurs in multiple interactive formats, which results in the creation of more investment opportunities and 'investor ready' companies. Wray (2012) also argues that in the North East of England, public, regional venture capital funds (RVCFs), although traditionally seen as a means to mitigate the equity gap (Harrison and Mason 2000a; Klagge and Martin 2005; Sunley et al. 2005), actually reproduce a territorial fixation of VC, which arguably is at the core of the region's VC problem. Hence, from a relational perspective on VC investment creation, both a fixation on territories and a fixation on either supply or demand seem problematic.

Classical venture capitalists in the narrow sense, i.e. institutionalized, yet private funds, are complemented by other types of investors in the marketplace. These alternative investors differ in their approaches to investment and the associated practices. One of the most frequently discussed types of alternative investor is the 'business angel' (Harrison and Mason 2000b). The term refers to individuals who invest their private fortunes in start-up companies and have a hands-on approach to providing management support. The added value associated with business angels is the experience, guidance and mentoring they provide to entrepreneurs.

There is ample evidence that business angels differ from classic venture capitalists. The sums invested by business angels tend to be smaller than investment rounds from VC funds. Furthermore, in the absence of a formal organizational structure, business angels tend to be less visible than VC funds. Establishing relationships between business angels and portfolio companies therefore depends more strongly on existent personal networks. In assessing potential investments, business angels tend to follow a more informal procedure and to put a stronger emphasis on 'soft' factors (Paul, Whittam and Wyper 2007). Compared to VC funds, business angels try to fixate fewer aspects of the investment relationship ex ante, during contract negotiations. Instead, they tend to follow an 'incomplete contracts' approach and negotiate a greater share of mutual obligations and agreements ex post, i.e. when the investment is sealed (van Osnabrugge 2000).

Regarding their relationship to venture capitalists, different lines of argument are followed. Harrison and Mason (2000b) found that there is substantial complementarity between venture capital funds and business angels or private investors, especially with venture capitalists who focus on seed funding. These include co-investments in the same firm, sequential investments in the same firm, deal-referring (i.e. exchange of recommendations regarding potential investment deals) and investment by private investors in VC funds. Co-investment appears to be the most common complementarity overall, although venture capitalists rate deal-referral as the most important form. Business angels appear to be particularly important for entrepreneurs in the earliest stage of their company founding. According to Madill et al. (2005) that investment by business angels is a very relevant determinant of consequent venture capital investment. Business angels help entrepreneurs to become attractive to more formalized forms of investment.

Angel investment is often discussed for its potential to close regional equity gaps. However, there is severe doubt about private investment's ability to close funding gaps. Avdeitchikova (2009) as well as Wallisch (2009) studied the spatial structures of the informal VC markets in Sweden and Germany respectively. Both find that informal investors tend to be strongly concentrated in metropolitan regions, according to Avdeitchikova even more strongly than regular VC. In the case of Sweden, there is even a reallocation of funds from peripheral to more central regions by informal investors. In sum, business angels can be understood as a very important and complementary part of the VC ecology, which strongly contributes to the creation of investment opportunities in dynamic regions.

A final type of complementary or alternative investment is corporate venture capitalism (CVC). In CVC, large corporations set up VC funds in order to invest in technology-oriented start-up companies. In doing so, they pursue both financial and technological aims. According to Weber and Weber (2007), CVC is a way to engage with and foster radical innovation. They employ the concept of 'relational fit' – a combination of social capital and knowledge-relatedness – to analyse how large corporations and biotechnology start-ups find each other through CVC.

Finally, variegated logics of VC investment cannot only be the result of varieties of investor types, but also of change over time. Szyliowicz and Madsen (2013) distinguish three investment logics: 'investing to build', 'investing for gain' and 'hybrid'. In the first logic, investors prioritize the creation of successful companies with economically sustainable business models, while still realizing a profit. In the second logic, maximizing financial returns is prioritized over the building of companies. In the third logic, both elements are combined. Studying the field of institutional VC over a 30-year period, Szyliowicz and Madsen conclude that all logics can co-exist without threatening each other. However, historic shifts occur. In US venture capitalism, the logic of 'investing to build' prevailed in the 1970s, while in the 1980s, 'investing for gain' was more dominant. In the 1990s, a renaissance of 'investing to build' was observable.

In sum, VC investors as the most visible representatives of financialization in innovation processes can be characterized in the following way: Their investment activity is spatially selective, building on opportunity creation, but also partially co-creating opportunities in environments of dense and diverse interactions of innovation actors. They enact relational distance to founders, enforcing norms and situated performances of professional business conduct. Their modus of relational work is best understood as a product of their own, often precarious relational position between institutional investors and exit markets. They can be expected to guide innovation processes in a particular way, reinforcing existent relations and spatial hierarchies. However, they are complemented by proximate but slightly different models, such as angel investment, corporate VC investment and public VC investment. The interplay between these diverse forms of capital investment in innovation processes is less predictable than a view on 'classic' VC alone would suggest.

1.6 Reconstructing innovation processes with innovation biographies

To make the time-spatial unfolding of individual innovation processes observable, a case study design was chosen. Each case studied represents one innovation process. In economic geography, process-oriented research concepts have been in use for some time (Hautala and Jauhiainen 2014; Ibert and Thiel 2009). The most common method of collecting data on individual innovations is the 'innovation biography' (Butzin and Widmaier 2016). In innovation biographies, innovations are identified ex post. Then, using qualitative interviews with participants, documents, visual material, ethnographic observations and other sources, the emergence and course of the respective innovation are traced and reconstructed. The method has the advantage that the relational and spatial contingencies of innovation can be observed – in contrast to methods which, for example, only focus on innovations' diffusion based on publications and patent citations (see for example Fritsch and Medrano 2010). As a result, the context sensitivity of innovations as well as the influence of existing relations and their role in search processes become visible, affording a more realistic and less objectivist view on innovation. Elements of both path dependency and path creation can be discovered through innovation biographies, making them a very holistic instrument.

In terms of geography, innovation biography research replaces the predominant focus on territories and territorial structure evolution in the field of 'territorial innovation systems' research with an open approach to time and space, relationality and materiality. The unfolding of innovations is traced across territorial borders. Social relationships, which occur in an innovation process, are observed and qualified as they are encountered – proximate or distant, productive, adversarial or ambiguous (Ibert and Müller 2015). Likewise, innovations are reconstructed through time and independently of institutionally defined time frames such as programme and evaluation periods, patent durations, changes to the legal framework or regional strategy formulation processes. Yet, naturally, this type of institutional time can be studied for its actual role in innovation processes – as a source of opportunity windows, as an impediment or in fact as both.

The dominant conceptualization of space is a topological one: places, either as loci of permanent interaction and co-location or as loci of temporary co-presence, are studied with respect to their specific, individual qualities. The latter are rooted in their historic evolution, multitude of socially ascribed meanings and the material 'thrown together' (Aitken 2010) presence of various actors. This conceptualization of space is rather the opposite of a territorial spatiality, which invokes notions of homogeneity and a non-material container space. However, territorial institutions and borders, just like institutional time frames, can also be observed if they occur as relevant factors in innovation processes and can be studied for their influence.

A large number of studies using innovation biographies are highly inductive and empirically driven (see Bruns et al. 2008; Butzin and Rehfeld 2008, 2012). Other studies use innovation biographies to analyse properties of innovation processes in a more theory-led perspective (Ibert, Müller and Stein 2014). In this study, the innovation biography is a tool to obtain data on innovation processes following an explicitly dynamic and non-territorial conceptual framework. The question of how innovations can be accessed in a linear way is relevant (Balconi, Brusoni and Orsenigo 2010). Once an individual innovation is identified, defined and delineated, the events which accompanied its emergence should, in theory, be traceable as a linear chain. However, when a process is reconstructed in hindsight, the ways in which events were ordered in time and in which they related to each other are always subject to ex-post rationalizations and narrative dramatizations. The data acquired in this way require an interpretative approach and cannot purely be treated as facts.

Accordingly, the interviews, which are conducted to create innovation biographies, are a combination of narrative biographic interviews (Fuchs-Heinritz 2009) and expert interviews (Gläser and Laudel 2010). At the beginning of data gathering, an initial respondent is identified, approached and interviewed. The subsequent interviews are initiated and conducted based on the results – which include names of other participants and potential respondents as well as recommendations in both directions – of this first interview ('snowballing', see Miles and Huberman 1996). This selective interdependence may produce a degree of cognitive path dependency or cumulated biases. However, having several interdependent interviews per case also provides an opportunity to counter bias.

In addition to bias brought in by individual respondents, researching innovations with innovation biographies is afflicted with the possibility of selection bias, or more specifically success bias (Collier and Mahoney 1996). Since innovation processes are reconstructed ex-post, failed innovations do not occur in case studies. It is well possible that action dynamics encountered in an innovation process are attributed a constitutive or supportive function by participants, while in fact luck and coincidence or other external, unswayable or even unrecognizable forces played a much bigger role. Two measures can counter this risk. Firstly, while collecting innovation biographies, innovation processes are not to be understood as smooth, mono-directional or even deterministic phenomena. Partial failures, break-ups, dead ends, changes of direction, re-contextualizations, phases of running idle – all these kinds of events can be observed within an innovation biography, thus accounting for the possibility of failure. And secondly, case studies can be used to triangulate each other. If certain patterns emerge repeatedly, they are likely to be relevant.

An emphasis was put on the role of finance in the process. However, this role was not addressed in an isolated manner, but in the context of other actions and logics. Therefore, investors and their relations to and interactions with other innovation actors shall receive prominent, but not exclusive attention. To have a clear starting point, I identified incidences of new product development occurring in or with German biotechnology companies, which involved capital investment activities in Germany. Hence, to appear as a potential case to study, the act of establishing a financing relationship between investors and product developers had to take place (predominantly) in Germany, and there had to be a clear indication that the products under development belonged to the field of 'biotechnology'. However, the cases were not meant to represent any concisely defined structural entity such as 'the green biotechnology industry in Germany' or 'the bio-pharmaceutical innovation system in Germany'.

Instead, the case identification process served to create starting points for a) a backward tracing of innovation processes and b) a tracing of the associations, but also separations of entities occurring in innovation-related work through action. In the categorization of Yin (2014; see Table 1.2), the chosen approach is a multi-case-study approach. Yin further differentiates case study research into embedded (each case study represents multiple objects of analysis) and holistic (each case study represents one object of analysis) case-study approaches.

The first person to be approached for an interview within a case study was typically the CSO (chief scientific officer) or CEO (chief executive officer) of a biotechnology company which offered or developed a new product. This choice was made because occupants of such functions had the most visible associations to innovation processes. In the case of CEOs, they also had first-hand experience in dealing with investors. The approach of a potential interview partner was preceded by desktop research about the respective company and innovation.

The defining criterion for an investor was that funds were allocated with the expectation of a return, that it was done under a subjective perception of conscious risk-taking, that some form of ownership (either of a company or of exploitation rights) was exchanged and that the expected future value created by the engagement in the investor's subjective perception at least was equal to the risk. There was no specification regarding the expected profit, the legal form of investment or the possible combination of financial valuation with other forms of value appreciation. Hence, a strategic partnership also qualified as an investment, since it was experienced as such by participants. The number of investors interviewed varied between one and three per case.

Interview partners were approached by letter, telephone or email. If they agreed to participate in an interview, face-to-face meetings were scheduled whenever possible. If respondents did not agree to a face-to-face interview, if no appointment could be made or if respondents were located very far away (e.g. in the United States), interviews were conducted by phone. In total, 38 interviews were conducted, five of these by phone. Interviews were audio-

Table 1.2 Overview of the case studies

Case	Organic mechanism addressed	Enabling technology	Application context	Duration
1 ENCAPSULATION	Encapsulating active ingredients	Chemical physics; agitation technology	Delivering pharmaceutical compounds	Mid-1990s to end of data collection
2 SYNTHESIS	tailoring peptide and protein molecules and observing biochemical interactions	Process automation/robotics	Pharmacological and immunological research	Mid-1990s to end 2000s
3 GENE FUNCTION	Monitoring impact of genes on cellular metabolisms	Gas chromatography, robotics, bioinformatics	Plant breeding (including GMO); discovery of pharmaceutical active ingredients	Mid-1990s to end 2000s
4 BIOMARKER	Observing DNA methylation patterns in the blood	Biochemistry, bioinformatics	Cancer diagnostics	Late 1990s to end 2000s
5 AUTOIMMUNE	Intervening in antigene-receptor binding mechanisms of the immune system	X-ray structural analysis	Autoimmune disorder therapy	Mid-2000s to end of data collection
6 STEM CELL	Cultivating non-embryonal human stem cells and making them produce human organic molecules	DNA technologies	Drug development (organic molecules)	Early 2000s to end of data collection
7 CANCER IMMUNE	Activating immune responses	DNA and Viral vector technologies	Cancer therapy	Mid-2000s to end of data collection
8 NEURON	Stimulation of receptors in neurons	Organic chemistry	Neurodegenerative disease therapy	Mid-2000s to end of data collection

Source: own design

recorded (except in one case following the wish of the respondent) and later transcribed verbatim.

The interviews conducted dealt with the innovation processes themselves and the ways in which actors got involved in them, affected them and were affected by them. The first interview of every case study was centred on the innovation process as a whole. Later interviews were more structured and specific. While following a common rationale, interviews with the broadly defined main types of actors (more investment-oriented or more science and technology-oriented) were conducted on the basis of specialized outlines. These were organized in two main parts, addressing a) the ways in which respondents described their own identities, positions vis-á-vis relational orders and everyday practices, and b) the ways in which they engaged with the innovation process studied, how they affected it and were affected by it.

In total, eight innovation biographies were collected as case studies (Yin 2014). This set of case studies can be characterized in the following way: the cases are diverse in terms of technologies, envisioned markets, emerging business models and forms of investment involved. Rather than being representative of German biotechnology according to predefined standards, they open up paths of access to the numerous internal differentiations which make German biotechnology.

One of these differentiations is the one between red and green biotechnology. The former refers to applications of biotechnology in medicine, that is, in therapy or diagnosis. The latter refers to applications in the agricultural and environmental field. A number of further colour denominations exist, such as white or grey biotechnology (related to chemical process technology) or blue biotechnology (applications related to maritime ecosystems). Red biotechnology is considered by far the largest field in terms of its market potential. Most German biotechnology companies have a connection to this field. Although I did not explicitly search for red biotechnology innovations, seven of eight case studies are clearly 'red biotechnology' innovations and one is clearly 'green'.

However, it is noteworthy that, while most of the cases have a very obvious linkage to medical therapy or diagnostics, this boundary is also subject to continuous renegotiation within the cases. In one case, an application in diagnostics was developed. Yet, scientists in this case study stressed that the underlying technology had been developed in a research field with an exceptional degree of overlap, commonality and cooperation between 'red' and 'green' researchers. They also stressed that the boundary between red and green biotechnology is not a consequence of 'objective' necessity, but of the closure tendencies of technology-related social circles. In another case, a technology was developed with an application in the agricultural field in mind. After successfully penetrating this market, the actors involved decided to establish an additional application in the medical field.

The cases relate in a similar manner to the concept of business models. German biotechnology companies are constantly searching for viable business models which allow them to attract investors and position themselves in markets. However, a business model is not necessarily a constant in an innovation process. Innovation processes cross company boundaries; hence, they can touch various business models simultaneously or sequentially. In the cases, the relationship between innovation processes and business models was under constant negotiation.

The cases are open-ended and diverse in terms of their times of emergence and their maturity when they were first observed. Finding entry points into a biotechnology innovation process is a difficult endeavour for one particular reason: Very often, biotechnology innovations are not marketed by designated biotechnology companies, but licensed to pharmaceutical and other industrial corporations. They are then marketed under a brand name which renders the biotechnological origins of the respective innovation invisible. Vice versa, as long as a biotechnological innovation is still visible as such, it has very often not reached market penetration. Additionally, both product development and market entry are drawn out, complex processes which unfold gradually. Having identified a fully completed innovation process, a researcher might find herself in a position in which the actor-network which had supported the innovation process has dissolved and the origins of the innovation are virtually untraceable.

Therefore, most case studies represent innovation processes, which were caught in the process of market entry. This can mean, for example, that an innovative pharmaceutical product has reached formal admission in Europe, but not yet in the United States. The latest stage, in which an innovation was initially observed, was a state of advanced market penetration. The earliest stage, in which an innovation was initially observed, was the beginning of clinical testing. Since the period of data collection spanned over roughly three years (mid-2010 to mid-2013), during which phases of intensive field work alternated with phases of preliminary analysis, all innovation processes were accompanied and occasionally revisited over a longer period. By observing company websites and press releases, I kept track of the innovations until mid-2015.

The innovation processes observed turned out to take a very long time, even by the standard of the often-cited long development periods in the pharmaceutical industry (10 to 15 years). If the first time at which innovators became aware that they were working on a specific innovation project (the beginning of the validation phase in Ibert, Müller and Stein 2014) is taken as a reference point, some of the innovation processes observed took more than 20 years. The earliest cases began in the early 1990s. The latest cases observed began in the mid-2000s. Very roughly, the cases can be divided into two groups: the early cases (beginning roughly between 1990 and 2000) and the late cases (beginning roughly between 2000 and 2005). This spread has two

Table 1.3 Overview of interviews conducted

Case	Interview no.	Position	Place	Mode
1	1–1	Founder/CEO	Berlin	Face-to-face
	1–2	Product development executive	Paris	Phone
	1–3	Cluster manager/investment manager	Munich	Face-to-face
	1–4	Investment manager	Berlin	Face-to-face
	1–5	CSO	Berlin	Face-to-face
2	2–1	Founder/CEO	Berlin	Face-to-face
	2–2	Scientist	Braunschweig	Face-to-face
	2–3	Pensioner, former product development executive	Berlin	Face-to-face
	2–4	Engineer	Cologne	Face-to-face
	2–5	CSO	Berlin	Face-to-face
3	3–1	CEO	Berlin	Face-to-face
	3–2	Scientist	Potsdam	Face-to-face
	3–3	Former co-founder, product development executive	Ludwigshafen	Phone
4	4–1	CSO	Berlin	Face-to-face
	4–2	Former co-founder	Seattle	Phone
	4–3	CEO	Berlin	Face-to-face
	4–4	Co-founder	Seattle	Phone
	4–5	CFO/investment manager		Face-to-face
	4–6	Scientist	Berlin	Face-to-face
	4–7	Investment manager	Munich	Phone
5	5–1	CEO	Munich	Face-to-face
	5–2	Scientist	Munich	Face-to-face
	5–3	Investment manager	Munich	Face-to-face
	5–4	Investment manager	Munich	Phone
	5–5	Investment manager	Munich	Face-to-face
6	6–1	CEO	Cologne	Face-to-face
	6–2	Co-founder/scientist	Cologne	Face-to-face
	6–3	Investment manager	Berlin	Face-to-face
	6–4	Investment manager	Bonn	Phone
	6–5	Investment manager	Cologne	Phone
7	7–1	Founder/CEO	Hamburg	Face-to-face
	7–2	Private investor	Stuttgart	Phone

Table 1.3 (Cont.)

Case	Interview no.	Position	Place	Mode
8	8–1	Founder/CEO	Mainz	Phone
	8–2	Investment manager	Bonn (interview Berlin)	Face-to-face
exp	e-1	Cluster manager	Berlin	Face-to-face
	e-2	Trade association leader	Berlin	Face-to-face
	e-3	CEO	Berlin	Face-to-face
	e-4	Investment executive	Darmstadt	Phone

Source: own design

consequences. Firstly, innovation processes were observed during different stages of their unfolding: the early cases were more complete, but associated with greater difficulty to obtain a reliable picture of their origins. The later cases were less complete, but afforded more opportunities for live observations of crucial path choices.

The cases are consistent with regard to the data gathering strategy. However, the number of successfully completed interviews varied somewhat. Two out of eight case studies comprise two interviews each; one case comprises three. Regardless of the small number of interviews, these cases were left in the sample. Vice versa, one case comprises seven interviews, as it covers a very wide range of places, historic phases, action settings and action logics, and thus is highly illuminating for the ecology as a whole. Table 1.3 provides an overview of all interviews conducted in the course of my empirical investigation. The column labelled 'position' reflects the position held by the interviewee at the time of interviewing, while the column labelled 'place' refers to the working place of the interviewee, which usually was also the place of the interview. The four rows labelled 'exp' refer to interviews with field experts, who, as pointed out before, to varying degrees also have relationships to individual cases.

The interviews within each case study were used to triangulate each other. Along with the ethnographic elements and additionally analysed documents, this triangulation was primarily used to arrive at more precise and weighed pictures of events. This step did not serve to invalidate statements made during interviews. Throughout the material respondents commented on other actors, their logics and their actions. They drew boundaries and invoked identities for themselves and others. The cases in which such statements supported each other on a more abstract, interpretative level (leading, for example, to the finding that two particular logics were regularly brought in fierce confrontation), by far outweighed the cases of unintelligible contradictions.

The interpretation of interviews is closely connected to considerations of research ethics. In order to protect the individuals who contributed to this study by participating in an interview, all names of persons and companies involved in the cases were made anonymous. However, to the inclined reader who is either very familiar with the subject matter or invests a degree of research, it will be possible to identify the cases studied and sometimes even the individuals. Biotechnology in Germany is a very small field, and biotechnology-related venture investment an even smaller one. Many respondents stated that they knew most other people in 'the community'. While it is difficult to verify these claims, I was able to verify that it is indeed very easy to grasp the 'who is who' in the field by studying attendance and speakers lists at the most relevant national trade events. Respondents participated in interviews in the knowledge that, with some effort, they could be identified.

2 The German case

Structures and relational dynamics in biotech financing

2.1 Building venture capital in a coordinated economy

According to the literature on 'Varieties of Capitalism' (Hall and Soskice 2001), national institutional systems and inter-institutional relations produce specific outcomes in terms of innovation. In applying this premise to Germany, a range of presumable incompatibilities between the institutional environment and biotechnology become apparent (Casper, Lehrer and Soskice 2006). Indeed, Germany began developing independent biotechnology comparatively late, in the second half of the 1990s. Since then, there has been an almost constant growth, albeit with phases of very different dynamism. Still, the German biotechnology industry remains small compared to America's and compared to established industries in Germany.

Arguments of institutional difference appear to capture important reasons for Germany's lag in biotechnology as well as certain tendencies of specialization in the existing biotech industry. Germany is considered a 'coordinated market economy' in contrast to the more 'liberal market economies' of the United States and also the United Kingdom (ibid.). The first category is typically associated with incremental innovation whereas the latter is associated with more radical innovation. The emergence of biotechnology itself clearly qualifies as a radical innovation. Company growth financing in Germany is dominated by bank loans and financing through cash flow, i.e. existing product sales. Both mechanisms are inadequate for allocating large sums to technology developments with a highly uncertain outcome and in most cases in the absence of collateral. In Germany, the stock market plays a subordinate role in company financing, especially in SME financing, and in the allocation of household savings. Large-scale private pension funds, which could exert a substantial demand for stocks do not exist.

Other institutional factors are cited as well. In Germany's coordinated capitalism, a multitude of formally organized actors – government bodies on national (*Bund*) and state (*Land*) level, trade associations, unions, chambers of commerce – engage in negotiations and seek consensus regarding most matters of economic policy. They do so within an institutional framework, which sanctions consensus orientation and distributes responsibilities which in most other countries are held by either the government or market actors. Labour contracts are still, in their majority, long-term contracts with very little scope for hiring and firing. In this environment of consensus orientation and long-term, trust- and reciprocity-oriented relationships, it is argued, disruptive changes to the economic landscape are discouraged. Other areas of difference are bankruptcy law (placing greater liability on entrepreneurs in Germany than in the US), taxation (limiting compensations of gains with losses in equity investment) and patent law.

But the nation-state is not a territorial container for institutional continuity, but also a playing field for political strategy, action and initiative. As discussed before, government policies played an important role in the creation of a biotechnology industry in the United States. More recently, the role of governments and nation-states in innovation policy has generally received increased attention and appreciation (Mazzucato 2013). Specifically, nation-states have the possibility to unfold technology-oriented agendas which suit their institutional settings, their social needs and cultural preferences, but also their market and population sizes (Breznitz 2007; George, McGahan and Prabhu 2012; Wong 2011). Large countries like India, China and Germany can be seen in the context of the 'developmental state' – the state which consciously and effectively tackles development deficits (Adleberger 1999).

In this chapter I want to integrate these perspectives – institutional and territorial difference, inter-territorial interdependence and specialization as well as state agency – and add to them my own dynamic relational approach to account for more recent developments. This line of argument is partially a consequence of the extreme richness of the case study material in terms of the changing institutional contexts for biotechnology innovation and financing. In this dynamic ecology actors relate to other actors, create relationships and sever them, construct relational proximity and relational distance.

Canzler, Wentland and Simon (2011) argue that a national strategy to establish a new technological paradigm requires two key elements: a narrative of solving an imminent problem and stabilization through the creation of inter-organizational relations. There are considerable differences in the evaluation of German policy performance in pursuing this goal. Giesecke (2000) is highly critical of the German approach, which she describes as interventionist technology policy. According to her, instead of improving the general innovation ecology, subsidies were allocated to a technology in an isolated manner, mostly through research funding. Adleberger (1999) provides a rather more enthusiastic account of German biotechnology policy. She stresses the historical context and the institutional constraints of the German government which, in a federalist and coordinated system, has very limited power to impose new technologies. German governments have attempted to foster biotechnology through research funding since the mid-1970s. However, they faced a largely complacent industry and fierce opposition to gene technologies in the public and several *Länder* governments.

In the early 1980s three 'Gene Centres' were set up in West Berlin, Heidelberg and Munich. These centres represent an early attempt to enrol industrial corporations for biotechnological research investment. They were envisioned as first-rate, well-funded research facilities, which were to create a push to innovation dynamics. The centres were half publicly funded and half funded by the industry. Although the expected impact did not materialize, the centres would later become important nuclei. In the early 1990s, Germany was in the midst of its post-unification recession. A lagging behind in high-tech industries was universally felt. In addition, industrial corporations had begun engaging with biotechnology – by building subsidiaries or investing in start-ups in the United States. Corporations justified this move with the uncertain, partially adversarial and fragmented regulatory situation in Germany (ibid.).

In this situation, the Federal Ministry for Science and Education (BMBF) initiated the BioRegio contest. In this programme, regional initiatives were financially and administratively supported in the development of regional networking and consulting activities to initiate biotech entrepreneurialism and to formulate an appropriate strategy. Crucially, these initiatives had to focus on small-scale, more functional interaction spaces and not *Länder* territories. All participants were asked to submit their regional strategy proposals. The three best ones were to receive DM 50 million in support funds – which had to be co-financed – to further develop their respective clusters or networks. The prospect of such substantial funding put pressure on regional policy makers to support biotechnology (Adleberger 1999; Dohse 2000).

The regions of Munich, the Rhineland and Heidelberg emerged victorious from the competition. These regions had been comparatively strong before (two possessed Gene Centres), and were further strengthened through the allocated funds. More importantly, however, in many German regions, serious political and economic support for collaborative action across institutional boundaries had been mobilized. The BioRegio competition is therefore credited with being the key to the surge in biotechnological company founding and investment in Germany in the late 1990s (Dohse 2000; Müller 2002). Another important area of political initiative was the creation of a publicly supported high-tech-oriented venture capital (VC) sector.

As with German industry, the government tried to win the financial sector's commitment to high-technology ventures for some time. Again, these attempts remained fruitless for a considerable period. As early as 1975, the government set up a fund called the 'German Venture Capital Company' ('Deutsche Wagniskapitalgesellschaft') with combined funding from the private sector (banks) and the public sector. This fund was to invest in new technologies and young companies. However, due to a lack of experience in managing and investing in such firms, the fund suffered heavy losses, effectively discouraging the financial sector from further engagements with venture funding. A second attempt in the 1980s, although more selective and risk-conscious, failed as well.

Another attempt in the year 1989 was finally successful. In the new framework, entrepreneurs were required to find private-sector investors, whose investments were then co-financed to the same amount up to a maximum of DM 1 million. Additionally, the public side would cover the risk of the private investors. Two public institutions delivered the financing: 'Reconstruction Loan Corporation' ('Kreditanstalt für Wiederaufbau' or KfW), the preeminent national public financing and administration body for public economic support programmes, and 'Technology Equity Company' ('Technologie-Beteiligungs-Gesellschaft' or TBG), a subsidiary of another national public financing body. Hence, both acted as matching funds for private investment and also as risk insurers.

In 1996, the programme was largely confirmed, but complemented by an additional financing option for entrepreneurs: company founders could now use public funding to buy back company shares from private investors and thus gain more independence. The programme was considered successful as it finally attracted substantial commitment of the German financial sector to venture capitalism. In 1998, 60 per cent of all German VC was provided by German banks (Adleberger 1999). In the same year, over 300 biotechnology companies were founded and roughly DM 425 million invested in them. In 1997, the 'Neuer Markt' stock exchange was set up to allow young high-technology companies access to the stock market with substantially lower bureaucratic barriers.

Due to these institutional reforms and policy interventions, a biotechnology industry did develop, starting in the mid-1990s and gaining considerable momentum towards the end of the decade. The business models of German biotechnology companies and the way in which they went about innovating was considered by observers to reflect adaptive behaviour to the institutional environment. Casper and Kettler (2001) argue that German biotechnology companies in the late 1990s displayed considerable overspecialization in the field of platform technologies. In a platform business model, a specific technological competence is exploited by offering research and development services to industrial customers instead of developing own end-user products. This tendency was viewed in sharp contrast to the very high-risk American model of the biotechnological (in most cases biopharmaceutical) product company. Casper and Kettler draw on the institutional properties of Germany's 'coordinated market economy' as an explanation, specifically the desire for immediate cash flow and low risk as well as the necessity to maintain long-term employment relations – a task which is easier to accomplish in a service company than in a project-centred product company.

VC institutions and biotechnology business models did not emerge by challenging existent structures, but rather by growing in their shadow. Biotechnology companies developed strategies of institutional hybridization, that is, they learned to operate in old and new institutional environments and use both as resources and 'tool kits'. This strategy allowed them to gain greater degrees of entrepreneurial freedom than a strict interpretation of the varieties of capitalism approach would predict (Lange 2008). Another interpretation of the term 'hybridization' relates to hybrid business models. In order to benefit from the relative security and constant cash flow of a service business model and seize the opportunity for large profits from a successful product launch, companies combined both elements.

The financial crisis of 2000 and 2001 temporarily called into question the entire notion of venture-funded high-tech entrepreneurialism in Germany. This crisis, following the burst of the so called 'dotcom bubble' was widely understood as a crisis of VC, specifically of overenthusiastic and unrealistic investment in high-tech ventures. Likewise, the use of the stock market to finance young, innovative companies was considered a failed experiment in the aftermath of the burst of the dotcom bubble. In 2003, the 'Neuer Markt' stock exchange was closed. Although very few biotechnology companies had been stock market listed, the perception of an open initial public offering (IPO) window had been an important driver of both company and venture fund creation. This window was now closed (Schudy 2006). For roughly three years, the VC market for biotechnology was extremely weak. The highly expansive approach of biotech start-up financing via TBG matching funds produced massive losses to public coffers and also drew criticism for its indiscriminate distribution of funds.

Also in 2003, a gradual recovery began, including a gradual opening of the IPO window from 2004 onwards (see Figure 2.1), which closed again during the financial crisis starting in 2007. However, important shifts became visible in the mid-2000s; instead of investing small sums in very young biotechnology companies, VC investors increasingly and very selectively invested large sums in later stage developments. Schudy argues that this concentration of funding

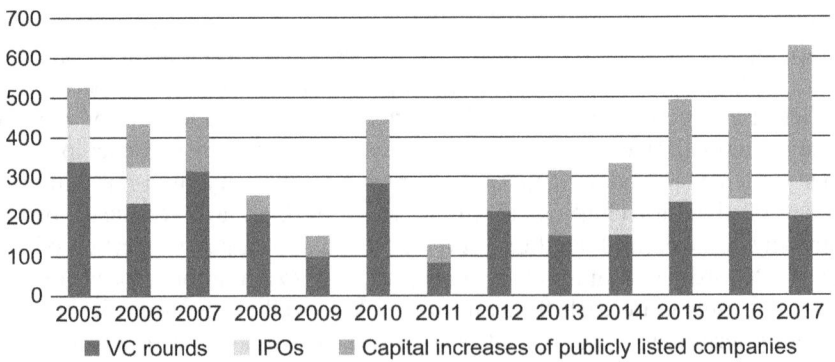

Figure 2.1 Capital raised by German biotechnology companies in million €
Source: Own design based on Ernst and Young (2010, 2015)

in few later stage developments increased problems of path dependency and lacking flexibility in biotechnology innovation, specifically red biotechnology, which had already characterized the sector. Platform-based and hybrid business models received more recognition in the aftermath of the 2000/2001 crisis. Furthermore, Schudy stresses the leading role of biotechnology companies which were created early on in the winning BioRegio clusters as examples of persistence and business model adaptation.

TBG investment was phased out in 2003. New public VC structures were established in the mid-2000s on the state and the national level, such as 'High-Tech Gründerfonds' (HTGF). So far, little is known about the impact of these new structures on biotechnology innovation.

Interestingly, the financial crisis of 2007/2008 was received differently from the previous one. While not being seen as a VC crisis, but rather a subprime lending crisis, the global financial crisis had a substantial impact on the VC sector (Birch 2016). Block and Sandner (2009) come to the conclusion that US venture financing rounds in 2008 were on average 20 per cent smaller than in the previous year. While early stage companies postponed expansion, later stage developments were unable to escape the weak IPO market and the downturn in VC funding. According to Mason (2009) the VC industry was oversized in advance of the crisis and different financing models (such as early exits in the form of trade sales) would reflect a more realistic investment practice after the crisis. Karberg (2009), specifically addressing biotech venture capitalism, also highlights the need for a healthy downsizing. He also emphasizes the smaller impact the crisis had on biotechnology companies in Germany compared to the United States. German biotechnology companies pursued more sustainable business models and were less dependent on a speculative capital market than their American counterparts. Equity investments in German biotechnology companies, including VC, have suffered a severe decline, however.

Investments until 2014 did not rise to the levels before the last crisis (see Figure 2.1). Furthermore, the recovery of biotechnology investment in Germany was slower than in the European average (Ernst and Young 2015). The volume of capital invested in biotechnology in Germany is substantially smaller than in, for instance, Switzerland and the UK. While in Germany roughly €120 million of VC were invested in biotechnology companies in 2014, in Switzerland roughly €300 million and in the UK roughly €450 million were invested (ibid.). Only in 2015 investment levels went back to the pre-crisis level, peaking (so far) in 2017 with €627 million (Ernst and Young 2018). The sum is spread across a wide range of investment models (public VC, private VC, family offices – see next section), geographical origins (German, European, American VC) as well as revenue from IPOs. The intermediate phases of decline after each crisis were extremely influential for the unfolding of the innovations studied here. After the crisis of 2008, financing relations in Germany were reconfigured in a highly specific way, the results of which are still palpable albeit less so than during the studied period.

2.2 Recent relational dynamics in German biotech VC

To offer a deeper insight into the qualitative changes faced by biotech-oriented venture capitalism, I will now turn to the relational embeddedness and relational strategies of VC firms as they lived through two major financial crises. The result is ambiguous. On the one hand, VC in the narrow sense has largely lost its ability to finance substantial biotechnology innovations in Germany, while at the same time becoming more entrenched in the national territory of Germany. Both developments contradict the alleged mobility of capital. On the other hand, new forms of VC have emerged since the mid-2000s and are now the cornerstones of biotechnology financing in Germany. To classic venture capitalists, three social environments and actor groups are particularly relevant, aside from their own portfolio firms (in this case biotechnology start-ups): the refinancing environment, i.e. institutional investors such as banks, insurances, pension funds, who invest money in VC funds; the exit market, particularly the stock market; and the field of potential co-investors. The volatility of the global capital markets in the respective time period, along with the properties of the German capital market environment affected the ways in which venture capitalists could move in this relational triangle, resulting in an increasing concentration of investment activity on the national territory.

Crucially, the institutional reforms to the capital market adopted in the 1990s created in market participants a sense that Germany was following the American model, a spreading notion of transatlantic commonality. Transatlantic VC companies emerged, which set up branch offices in several German and other European cities, such as 3i. During the technology boom of the 1990s, American investors were active in the German start-up scene and German VC companies ventured into the American market. German institutional investors such as large banks set up VC funds or invested in them. As stated before, the general optimism regarding the advances in the field of genomics additionally increased the positive attitude towards ambitious biotechnological business models such as biopharmaceutical product development. In this period, a high degree of relational proximity was felt both between investors and technology developers and between venture capitalists on both sides of the Atlantic. Venture capitalists were willing to invest in very young biotechnology start-up companies originating from basic research environments. In an optimistic assumption that the genomics wave would lead to a range of new medical therapies, based on the notion that genes solely determine organisms' properties, venture capitalists attempted to monopolize new research fields through patenting. Roughly simultaneously to the burst of the dotcom bubble in the capital market, these assumptions were found to be wrong.

With the first major financial crisis which hit the young German VC and technology start-up scene, latent differences between the German and the American institutional environments became apparent in a drastic way. This refers, for example, to German labour legislation:

In those years, we had frequent inquiries from the USA of course. We had investors who invested quite actively in the Munich area, roughly until the first financial crisis of the decade came. Then, all of a sudden, the Americans said, 'We have no interest anymore'. This is when they got to know our legislation, our labour law. They noticed when a company has difficulties, in the USA they can save it simply by firing 99% of the workforce. You can't do that here. I distinctly remember a meeting where an American investor said, 'I will never invest in Germany again'. This was a lesson to him, as it were. He said, 'In the United States, we could have saved this company. Here we cannot do anything'. The company then broke soon after. The employees were laid off; they immediately went to court and said, 'I want a one-year indemnity'. (Interview 1–3)

As one respondent pointed out, VC funds are closed funds. This means that a financial crisis does not immediately impact their ability to invest – the money is already raised and does not disappear. What was diminished in the financial crisis was a) the prospect of raising funds from institutional investors in the future to set up new funds and b) the prospect to successfully bring an investee company public and create revenue by selling stocks. While it remains unclear in how far such changes impact on a VC fund managers' investment choices (which always include a great deal of anticipation), it is obvious that the prospects in both arenas have a long-term impact on venture capitalists' ability and propensity to invest.

Respondents drew a sharp line between the internal workings of the German and the American capital markets. The American stock market was described as cyclical, but reliable. Boom-and-bust phases were said to alternate, creating periodically opening and closing 'IPO windows'. Private pension funds were said to be the most important players on the American stock market, creating a continuous demand for stocks. In Germany, on the other hand, every financial crisis is perceived as an existential threat to the stock market financing model. Only a small minority of German private households possess stocks. In the absence of a capital-market-based pension system, there is no stable demand for stocks other than the most established ones. As a consequence, the most natural exit path for venture capitalists is perceived as blocked.

In a similar way, the two markets are described as fundamentally different with regard to institutional investors' presence and readiness to invest in VC funds. Germany does not have a fund-based pension system, which is the backbone of institutional investment in many more capital-market-oriented economies. Additionally, legal constraints on the German side, such as banking regulation and tax regulations are considered adversarial to institutional investment in VC. These perceived differences cannot be discussed in detail. Here, the differences are relevant with regard to the ways in which they are perceived, dealt with and altered or confirmed in the relational practices of VC investment.

The question of how difference plays out in practice – as an obstacle or as an opportunity for complementarity – needs to be addressed. In theory, American and other European VC funds can always invest in German bio-technology companies. Furthermore, American (and other) institutional investors can invest in German VC funds. Such international investment relations are created and several respondents argued that such relations were indispensable to maintain some degree of investment capital supply to inno-vating German biotechnology companies. One representative of a European life science VC fund with a strong presence in Germany stated that less than 5 per cent of the managed investment sum stemmed from German investors and the majority came from Scandinavia and the Benelux countries.

However, the weakness of German institutional investment in VC funds as well as the absence of German demand for young-company stocks is seen as a major obstacle to more substantial international investment relationships. For once, there is a pronounced national bias on stock markets (Wójcik 2011), which means that there is a general preference, on the side of institutional investors, for the home market. This bias is confirmed by interview partners.

> US pension funds find it more comprehensible to invest in American companies. (Interview e–2)

In addition to this generic bias, there is an expectation of reciprocity. One VC representative explained that, to international pension fund managers, a sub-stantial co-investment by German banks and insurance companies would be a precondition for an investment in a Germany-oriented VC fund. As many stu-dies have shown (see Fritsch and Schilder 2006), VC investment consortia require at least one local investor who is intimately and culturally familiar with the investment and can be physically present on a regular basis. This means that international VC funds require a German partner to invest in a German com-pany. This principle seems to apply on a higher level, too. Creating a VC fund in Germany requires investment from German institutional investors; otherwise, international institutional investors increasingly refuse to participate.

In essence, both mechanisms represent considerations of reciprocity which are employed to reduce uncertainty in collaborative investment practices. They are complemented by a third principle. In the cases, on several occa-sions, individuals 'switched sides' from the investor to the investee side. The boundaries between the two sides are thus permeable. Several investors argued that having invested in a particular field or territory (for example, by being present in the US as an investor) is also a vehicle for creating new investment opportunities, both actively (as investor) or passively (to find investors). Venture capitalists have to perform both roles, investor and inves-tee. The multi-layered-ness of investment opportunity creation implies that attracting VC into a national territory is eased if investment capital directed at venture investments also flows out of said territory and there is, hence, a mutual, trans-territorial engagement.

From the early 2000s to the end of data collection, several occurrences indicate a territorial fragmentation of life-science-oriented venture capitalism. Firstly, respondents reported a diminishing number of VC funds which were actually active in life science-oriented venture investment in Germany. In fact, the existence of a German VC 'scene' in biotechnology was repeatedly denied.

A small number of funds are named as 'the last ones standing', such as Life Science Venture Partners (Munich/Amsterdam), Wellington Partners (Munich) and Creathor (Bad Homburg/Frankfurt). Each venture fund has only a limited time window to realize investments; hence, the number of available investors at any given time is still substantially smaller. Secondly, international life science VC companies massively reduced their presence in Germany, closing branch offices altogether or reducing them to one, usually located in Munich. Former transatlantic or European VC companies refocused their activities on the national home market, mostly in North America. Respondents cited 3i, TVM and Atlas Venture as examples.

Under these circumstances, it has become increasingly difficult to allocate substantial funds to biotechnology innovation processes. While until the early 2000s, life science VC funds managing €300 million or even €500 million were active in Germany, from the late 2000s, a VC fund was considered large if it managed €150 million to €200 million. To regular VC investors, the consequence is a different, more cautious attitude towards biotechnology. The investors interviewed stressed that there was no such thing as an ideal business model or model of innovation funding, as is often implied when the concept of 'hybridization' – the combination of product development with a cash-flow-generating service platform (Bindseil 2005) – is discussed.

The attitudes towards biopharmaceutical product developments were rather diverse. While some respondents found product development to be entirely unfeasible for venture capitalists due to its high cost, high risk of failure and the long development timelines, others still considered it a manageable endeavour. Yet, it was beyond dispute that since the early 2000s, venture capitalists have been engaging with biotechnological product developments at a later stage and demanding a higher level of proof and validity (Schudy 2006). In practice, this entails the completion of preclinical trials or the entry into clinical trials. It was furthermore commonly agreed that 'low-risk' and lower-cost business models such as diagnostics, medical technology and technology platforms have become more popular.

Many respondents stressed that on the capital-demand side, approaches to biotechnology had changed, too. More recent company founders are considered more experienced, professional and cautious than their predecessors in the 1990s.

> Many are Mittelstand. And [...] our founders have been in business for 15 years and resent the notion that they are trainer[1] people who have just come out of university, looking scrubby. They are real entrepreneurs and not at all set on being seen as 'trainer-type founders' and who rather say,

'I am a respectable Mittelstand entrepreneur and I want to be treated like one'. (Interview e–2)

A small number of repeating or even serial entrepreneurs have emerged. VC investors now treat previous entrepreneurial experience in biotechnology as a quality seal and sometimes an obligatory precondition for investments. The industry as a whole has matured. Companies from the earliest phase of bio-tech entrepreneurialism, such as MediGene and Evotec, have lived through several crises and adjusted their business models. They are now considered successful *Mittelstand*[2] companies.

The gradual retreat of classic venture capitalists from biotechnology in Germany was met by alternative movements and forms of investment from the mid-2000s. At the end of the data gathering period, these alternative forms appeared to be more influential and relevant to the financing of bio-technology innovations in Germany than conventional venture capitalism. These new forms bring with them a new territorial orientation – both as material entanglements and as cognitive spatialities. The argumentative start-ing point for describing the trend is the increased role of wealthy individuals and industrial families.

In the absence of a capital-based pension system and given the abstinence – still in effect towards the end of the 2010s – of most German households from the stock market, average individuals are enrolled as subjects of financialized circuits of accumulation (Sokol 2013) to a lesser extent than they are in more capital-market-driven economies. High-income individuals and specifically family owners of industry companies, on the other hand, are more inclined to engage in entrepreneurial finance. Managing and advancing the family for-tune combined with a philanthropic inclination can be a motivation to pursue start-up financing. For example, the largest holder of BMW shares (just below 50 per cent in 2014) is the Quandt family, arguably the richest family in Germany. Family heir Susanne Klatten is a renowned investor in German industrial corporations and noted for a long-term-oriented, *Mittelstand*-friendly investment style (Freitag and Student 2013). In 2011, she began set-ting up an investment fund focusing on start-up companies in the field of medical technology.

In the field of biotechnology, similar dynamics are observable. One example is SAP co-founder Dietmar Hopp, whose private fortune is estimated at around €7 billion. Next to engagement with a Bundesliga football club and a number of charity activities, Mr Hopp set up a VC family office, which is one of the largest VC investors in German biotechnology (Oberhauser-Aslan 2014). A variety of smaller VC family offices exist. However, in the studied cases, a different family office (which for many years accounted for double-digit percentages of all VC investment in German biotechnology) became influential. Two brothers from South West Germany had founded and mana-ged a leading German generics (non-patent protected drugs) company. The company was sold to a Swiss pharmaceutical corporation in 2006. Out of the proceeds estimated at around €6 billion, among other things, a VC family office (SOUTHWEST FAMILY VC) was set up, which manages a fund in the size range of €1 billion.

Family offices were a controversial topic among respondents. Unfortunately, no representative of a family office participated in an interview. Nevertheless, a picture could be formed based on other interviews and a media analysis. When asked about the differences between the family office model and other forms of venture capitalism – as in other comparisons of investment logics – a frequent answer was, all investors want the same thing – to make profit and not to lose money. Sometimes, this statement was a precursor to an admittance of dividing lines, which were in fact rather substantial. Furthermore, the statement 'we want what everybody else wants' sometimes appeared to be an invocation of commonality which served to legitimize participation in venture capitalism on equal footing with the 'natives', especially on the part of public VC investors.

Family venture capitalism is an independent investment model following its own logic of value appreciation and creation. Family investors are at liberty to impose their own time frames and return expectations instead of having them imposed by institutional investors. Another factor is fund size. A classic life science VC fund with €200 million under management was considered reasonably sized or even large by respondents. Managing €1 billion in combination with an essentially open-ended time horizon allows for a fundamentally different conceptualization of risk and risk distribution. This, of course, only applies to the very largest family offices. The third aspect is intrinsic motivation. Individual ambition, appreciation for entrepreneurialism and philanthropy meet in the action logics of family entrepreneurs. The associated conceptualization of value also appears to have a regional component.

The actions of the founders of SOUTHWEST FAMILY VC appear to support these assertions. As mentioned before, a part of the proceeds from the company sale was invested in a new life science basic research institute. Hailing from southwest Germany, in 2004, they also had invested in a regional bank and in 2014, substantially ramped up their investment.[3] With the creation of a biotechnology-oriented venture fund, a very idiosyncratic approach was pursued. At the end of data gathering, SOUTHWEST FAMILY VC had invested in 11 German biotechnology companies. Supporting German biotechnology instead of continuously 'looking to America' was a self-declared aim of SOUTHWEST FAMILY VC. A substantial part of the investee firms were drug development companies (see Chapter 4) which had been in an early development stage at the time of investment.

Probably the most distinguishing feature is the systemic approach of SOUTHWEST FAMILY VC. Classic VC funds create portfolios based on considerations of risk distribution. Hence, every investment is assigned a risk profile combined with other financial metrics, beyond which there is no interdependence in the portfolio (disentanglement, Callon 1998). Some life science funds specialize in particular markets or fields, such as neurodegenerative diseases. One respondent argued that this was in part a strategy to distinguish oneself from others in communication with institutional investors. Crucially, investments within a classic VC portfolio are *not* technologically or scientifically interdependent in the sense that they a) are complementary and build upon each other or b) deliberately follow different schools of thought on the same subject so as to hedge against failure.

SOUTHWEST FAMILY VC very deliberately pursues the first kind of interdependency. As several respondents stressed, the office's portfolio represents an attempt to build new bio-pharmaceutical corporations out of biotechnological start-up firms. Complementary technologic capabilities are thus acquired and linked. One very prominent example is a start-up from Mainz University, which was funded and strategically transformed into an umbrella organization for further acquisitions, including drug developing companies and platform-based service providers. Thus, new functionally integrated multi-site networks emerged within a national territory. Following the recovery of VC markets in the mid-2010s, this new integrated company received two international financing rounds so big (each one amounted to several hundred million Euros) that they were perceived as distorting the statistical picture of VC investment in German biotechnology (Ernst and Young 2017). Since the aim of SOUTHWEST FAMILY VC is in part to build new, profitable companies, this approach comes closest to the logic of 'investing to build' described by Szyliowicz and Madsen (2013). The approach does not apply to the entire portfolio, however.

Representatives of conventional venture capitalism argued that SOUTHWEST FAMILY VC usually 'did their own thing'. They did not seek collaborations with classic VC funds, who could neither match their investment volumes nor their temporal endurance. The responses to this independent approach varied. It was praised as visionary and effective by some, but it also met criticism, specifically with regard to the excluding attitude.

> The [*SOUTHWEST FAMILY VC*], yes, and they have such a lot of money, and that means also time, that they cannot be rushed. They act, I would say, alone. Indeed, in my view they are building, if you will, something like an empire. They pull it through alone and for themselves. (Interview 6–4)

One respondent was fundamentally critical about the possibility of building new biotechnology companies on a large scale and branded the underlying ambition as irrational enthusiasm. More generally, the activity of very large family offices was met with skepticism.

> I believe [name] has no idea what he is doing. He put a lot of money into this handful of companies he has got there, but I doubt whether he understands it. (Interview 6–4)

Family offices also change biotechnology-oriented venture capitalism indirectly, in addition to their own investment activity. Wealthy individuals and VC family offices increasingly invest in regular VC funds – an interdependency which was already described with regard to the relation of business angels and venture capitalists (Harrison and Mason 2000b). The growing relevance of family-based investment for regular VC becomes visible if the fund creation histories of VC companies are taken into focus. In the cases studied, two regular VC companies appeared which displayed a funding shift from institutional investors to private individuals, family offices and also public banks over the course of two and three funds under management respectively. While the funds created in the late 1990s were larger and funded by institutional investors, later funds were smaller funded predominantly by the alternative investors named.

SOUTHWEST FAMILY VC frequently forms joint investment consortia with a VC fund which pursued a fundamentally new business model. This fund will be referred to as retail-finance-based VC or 'RETAIL VC'. SOUTHWEST FAMILY VC and RETAIL VC collaborate on a regular basis. Their logics were perceived as compatible by respondents. They are similar regarding the fund volume under management and the investment time horizons, and in both respects, differ from conventional VC in Germany. One respondent stated that RETAIL VC often acted as a representative of SOUTHWEST FAMILY VC in investment consortia. Hence both funds enact relational proximity based on fund size, investment time frame and preference for investments.

RETAIL VC emerged in the mid-2000s in Munich, and its creation can be considered a business model innovation, possibly even an instance of institutional entrepreneurship, as hitherto disparate institutional orders were recombined: VC and financial services for private households. The participants were a Munich retail finance company, which sold mainstream investment products like life insurance to private households via a nationwide distribution network, and a likewise Munich-based group of industrial experts who advised investment funds. Their co-location allowed overlaps of everyday action spaces. A meeting was brokered by a lawyer who had worked for both parties. One participant made the suggestion to combine the retail finance distribution network (consisting of several hundred sales agents operating from sales offices across the country) with a more ambitious investment agenda. Consequently, a company was created which sells VC style investments to private households.

In essence, a sequence of closed funds was set up. Two forms of funds exist: either a minimum amount of €6,000 can be invested for ten years, or larger sums, such as €50,000 can be invested gradually over a longer period (such as 20 years). At the end of data gathering, 13 funds were in operation. A new fund was established every 18 months. In total, around €800 million, stemming from roughly 50,000 customers, were under management – a financial volume comparable to that of SOUTHWEST FAMILY VC and vastly larger than an average conventional life science VC fund. The investments were sold with an envisioned return similar to that of a VC fund. Likewise, similar to VC investments is the – openly communicated – risk of losing the investment.

This construct has profound implications for the company's investment practice. Most importantly, fundraising and investment management are separate tasks. In a conventional VC company, investment managers have to go through a cycle with every individual fund: approaching institutional investors, raising funds, initiating a 'deal flow' (i.e. investing in companies), managing the investments, disinvestments ('exit'), realizing returns and terminating the fund are all tasks which are organized sequentially over a ten-year period and which are performed by investment managers (see Klagge and Peter 2009). In practice, this allows for an investment duration of roughly five, in exceptional cases seven, years. In the case of RETAIL VC, the pre-existing retail network with professional fundraisers permanently approaches potential clients and raises funds. The company's aim is to maintain a stable pace of new fund creations and (in the future) payouts. On the other side, investment managers are permanently in the process of selecting companies for investment, managing the investments and performing exits. Both activities are decoupled from each other and run simultaneously.

The rather unusual relationship to individual households, who are enrolled as quasi-VC investors, lends a particular territorial quality to the financing model. Firstly, the pre-existing retail network with its branch offices and sales representatives operates nationally – as is to be expected from a sales network which sells products like life insurances. The envisioned target group consists of wealthy households with minimum liquid assets of €100,000. A company representative explained that, due to the products' high-risk nature, customers are advised to invest a maximum of 5 to 15 per cent of their liquid assets.

Potential customers are interested using two key arguments: a monetary argument, which highlights the earning opportunities in comparison with widely used savings vehicles like savings accounts and government bonds; and an idealistic argument, which highlights the perspective to invest in socially desirable entrepreneurial activities. This approach entails a secondary territoriality on the investment side. Potential customers are assured that their investment flows into company funding almost exclusively in Germany (and also Austria). Hence, a form of economic patriotism is capitalized on as an intrinsic motivation to invest.

> Firstly, because we can plausibly explain to the private investor that we reinvest the money, which he earned and on which he paid taxes, in his economy. People feel connected to their home; this is a very important factor. Before we invest in shopping centres in Bucharest or in sides of pork in Szechuan or some such thing … People have a connection. That's why this is an important sales argument. (Interview 5–3)

In fact, RETAIL VC's investments are heavily concentrated in South and South West Germany. Like SOUTHWEST FAMILY VC and unlike the remainder of life science-oriented VC in Germany, RETAIL VC has a strong focus on drug development. In addition to the territorial focus, the fields of investment are a relevant subject of customer communication. RETAIL VC focuses on innovating SMEs and start-ups. The branches and technology fields of investment (IT, environmental technologies, 'health') are chosen and delineated in a way that makes them presentable as both innovative and socially desirable.

> And apart from that there is something essential, and that is the fact that it touches people, because it has to do with health and life quality. We distinguish three types of biotechnology: red, white and green. Green, we don't do, only red and white. And red biotechnology is of course the one related to health. We don't sell investments in biotechnology to private investors, but in health. (Interview 5–3)

These delineations represent an active engagement with national social orders. Polls produced representations of society which indicate that 'the Germans' appreciate the use of biotechnology for medical purposes, but reject agricultural biotechnology (Kutter 2014). Consequently, agricultural applications of biotechnology, specifically GMO development, are excluded from RETAIL VC's portfolio.

As a representative from an investee company explained, RETAIL VC attempt to make the entrepreneurial activities in which funds are invested visible and palpable to customers. They also appeal to – arguably – specifically German orders of worth, in which more traditional forms of financial and entrepreneurial risk-taking, like lending, are appreciated whereas the speculative nature of the stock market is rejected. The approach resembles other historic examples (Garon 2012; Zelizer 1978) of attempts to build new premises for reconciling forms of financial investment (here: VC) with conflicting societal values. This form of relational work has its own spatiality.

> They have private customers who are frustrated because stocks keep going up and down. And they say, 'We invest in ethical companies: health market, medical technologies, environmental protection and modern technologies, information technologies … all things which serve humanity'. Everybody knows what the companies do. You can watch them. I always have to give presentations at their sales events. 'Meet the CEO' [*literally 'Geschäftsführer zum Anfassen' or 'CEO to touch'*]. (Interview 5–1)

The key to this model is a new interpretation of the private household as a financial subject in the context of both global and national financial circuits. Although German households are subject to tendencies of excessive indebtedness and thus arguably to an exploitative financial subjectification (Sokol 2013), saving is still the more dominant and more morally appreciated financial practice (Garon 2012). Puzzlingly, even in an era of effectively zero interest, government bonds and bank or *Sparkassen* savings accounts are still the most popular financial products. In this situation, RETAIL VC attempt to enrol the private household as a risk-taking investor, apparently successfully (although no payout has yet taken place). According to their own reckoning, the typical private household investor and RETAIL VC customer is male, over 40, wealthy, Western or southern German and has an entrepreneurial or technical background.

However, this role definition is challenged. The key point of attack is the question of whether private individuals are able to and should be put in the position to assess and shoulder the risk of venture investment. A representative of public bank KfW stated that his organization would refrain from entering investment consortia with RETAIL VC based on the potential of such financial practices to be scandalized in the case of failure.

> There is a significant default risk for the individual investor. And they don't have just €50,000 deposits; there are small investor funds, too. And I would never do VC investments using small investor funds, never. [...] The small investor is a very hot potato; because when Aunt Emma[4] loses her €5,000 because some company in Regensburg doesn't reach its goals [...] there will be a lot of clamour in the Republic. And when it turns out KfW[5] was in on it, oh boy! This is why we tend to stay away from these small investor funds of [RETAIL VC]. (Interview 6–4)

The role of the private household as an investor is also subject to efforts in financial and banking regulation. In the aftermath of the financial crisis starting in 2007, such investment activities which were deemed 'risky', specifically on the side of banks serving private customers, were put under increased scrutiny and also limited in quantitative terms. In turn, the Basel III banking regulations were heavily criticized by representatives of the German biotechnology industry, arguing that German *Mittelstand* loans, which now were deemed 'risky', had been solid as a rock during the crisis, whereas government bonds, now privileged by regulation, had been the source of substantial trouble, specifically in the context of the Euro crisis. In legal terms, RETAIL VC is a form of alternative (non-bank) investment fund management. In order to protect private customers from risk, such activities were regulated by the EU's Alternative Investment Fund Managers Directive of 2011. It was pointed out that the national implementation of this directive, if passed in the way it was initially planned would have been so strict that biotechnology-oriented investment in Germany would have suffered a dramatic loss in financing.

> We currently have the AIF directive which came from Brussels. If it had been implemented in the way our government initially suggested, the [RETAIL VC] funds would have been dead. They are an important column of financing and we fought for it to continue. (Interview 36)

RETAIL VC's own investment activities in biotechnology can be characterized as selective, organic and sensitive to pre-existent relational and spatial entanglements. The company's representative stressed that a profitable exit was RETAIL VC's aim, as is the case in every other VC company.

> We always think from the end. We are interested in the answers to the five Ws: who will pay how much, for what, when, why? (Interview 5–3)

However, it became apparent that RETAIL VC differs substantially from other VC companies in Germany regarding its appreciation of Biotechnology. The general skepticism of venture capitalists regarding biopharmaceutical drug development was not shared by RETAIL VC. The reason given was that the notion of 'high risk' is not an absolute, but rather depends on the investment business model. RETAIL VC can invest substantially bigger funds than an average VC company, which means that unplanned cost increases are manageable. The investment time frame is also more flexible. While some exits are realized after a very short time, the company can stay invested substantially longer than the average VC in other cases. In contrast to conventional venture capitalism, RETAIL VC can a) use entire fund durations for investments and b) perform crossover investments between funds.

RETAIL VC receives business plans and investment proposals via multiple sources: electronic communication, business plan competitions, conferences and publications.

> On the website people can download something and send something in, our colleagues walk around like truffle pigs, they take that as a compliment, they are continuously at conferences, business plan competitions, scientific events, founder exhibitions, trade fairs and, and, and. They read literature, really look at everything, talk to people where the interesting ideas are. They really do that actively. (Interview 5–3)

According to their own assessment, RETAIL VC receives around 98 per cent (in numbers roughly 1,500 per year) of all investment proposals above €1million 'floating around' in Germany. Roughly 80 per cent of these are dismissed immediately. The key criteria for following up on an investment proposal (that is, discussing it internally and then inviting the entrepreneurs to present) are a) a viable business case and b) a competent management team. The early stages of an assessment are performed in-house and strongly based on experience and instinct. As in other VC firms studied, the investment managers have first-hand entrepreneurial experience in the field in which they invest. The interview partner repeatedly employed the notion of a 'qualified feeling' to describe, what he and his colleagues searched for in situations of uncertainty. He argued that after having worked very intensively with both the managerial and the scientific and technical aspects of biotechnology for a long time, he had heuristic rules to assess proposals.

There are the classical risks of pharmaceutical development, regarding the molecule. You have to distinguish carefully between small molecules and biomolecules – antibodies or therapeutic proteins. And with the small molecules we are very careful, because they inherently have a whole range of activities which are always dangerous. These are unspecific interactions with other target structures in the body, reactive metabolites, metabolism and degradation difficult to control. From these we shy away. (Interview 5–3)

Crucially, the proposals which are forwarded by former collaboration partners were considered the most attractive.

We have a pretty good relationship network of people, of partners, who know what we are interested in and in what not. Surprisingly, via this route we only receive very few, highly qualified products. Most people find it difficult to assess what we like in the end and what we don't like. (Interview 5–3)

Also, a small number of entrepreneurs with a track record of good leadership were always considered attractive. Following the internal assessment, the network of current and former investee firms is used as a source of references regarding technical and scientific issues. Hence the relational space of previous investment engagements is a key source for the creation of future investment possibility spaces. More formal ways of technical evaluation (partly by external experts) and due diligence (finances, intellectual property situation, legal questions) occur comparatively late in the assessment process. As a consequence, previous relational entanglements and the associated decision heuristics of investment managers are extraordinarily influential in the selection of future investments. The possibility that this course of action produces cognitive lock-ins was highlighted by the interview partner.

It's a combination of gut feeling, experience, different perspectives, including different experiences among ourselves, certainly also prejudice. Of course, there are clinches. Let me give you an example. Somebody comes along and has a new method to detect metastasized tumor cells. It's an important question: is a tumor metastasizing, yes or no? We understand the pathways how metastasis happens, we know the key mechanisms, we know which methods exist to identify disseminated tumour cells, to accumulate them. We know the key markers and can judge fairly quickly: what is the business model? Do they have a new technology or do they also have a device? Is it a medical product or just a new procedure? If yes, can it be patented? Is it superior? [...] On what level is it superior? [...] Does it have better markers to identify the metastasizing cells, is it robust, is it easy to implement, can you use it on existing platforms, and can you, in the end, use it to justify a therapy decision or any other kind of decision, which represents any kind of value? And we can judge that fairly quickly. (Interview 5–3)

I understand this dependence on personal experience and intuition as an indication that in biopharmaceutical product development, there is no formalized metric which could reliably reduce uncertainty and complexity. A successful and profitable exit was described as the sole goal of each investment. The path towards this goal, however, was understood as a development which takes place individually in each investee company. While RETAIL VC shared the notion that pharmaceutical corporations represent the primary exit path, there was no generally pursued 'golden way' of approaching them. Instead, working intensively with management teams on situation specific issues was seen as the best way to prepare the best possible exit. As stated before, the duration of the investment is very flexible. Depending on the situation, an exit could have different consequences. While some companies are perceived as having the potential for becoming a 'centre of excellence' within a large corporation or even for staying independent companies, others are perceived as temporary project ecologies, which will dissolve after a product is launched. The Mainz-based biotechnology start-up, which SOUTHWEST FAMILY VC intended to build as a new biopharmaceutical company, and in which RETAIL VC had co-invested, was cited as an example for an explicitly company building-oriented investment.

As stated before, RETAIL VC frequently collaborates with SOUTHWEST FAMILY VC in investment consortia. Beyond that, the company cultivates a collaborative attitude towards other VC investors. Due to the small number and size of VC funds in German biotechnology, the main focus is on international, including transatlantic collaborations (substantiating the notion that the mobilization of substantial domestic funds is necessary for enrolling more distant partners). In this role, too, RETAIL VC appears to occupy a compensatory position for institutionally financed VC funds, which are not present in relevant size in Germany.

2.3 Creating relational proximity in public venture capital provision

After the perceived failure of public financing institutions in the dotcom crisis, new approaches emerged. The key idea herein is to attract private VC by creating relational proximity between private and public VC. Judging by size, the preeminent actor in this field is still KfW. KfW or Kreditanstalt für Wiederaufbau (Reconstruction Loan Corporation, whereby 'Reconstruction' refers to the organization's initial role in post-war West Germany) is the preeminent public financing body at the federal level. Quantitatively, it is the third biggest bank in Germany. It acts as a financier with a clearly political mission. This includes the formal administration of a large number of government funding schemes, but also the delivery of financial products aimed at activities which are deemed desirable, such as thermal insulation refitting of homes or solar power installations. KfW also invests in companies if their activities meet governmental priorities. It is best characterized as a political bank. It is supervised by the Federal Ministry of Economic Affairs (BMWi).

As discussed in above, along with TBG beginning in 1989 and in an expanded way since 1996, it provided matching funds for seed investments in biotechnology and other high-tech fields, and also small-scale seed funding and consulting for company founders. These measures were particularly effective in localities where technology and founding-oriented support structures existed, which were able to 'draw down' KfW and TBG funding (Sunley et al. 2005).

These changes can be explained with shifts in KfW's organization and policy. Following the burst of the dotcom bubble and the following financial crisis in 2000–2001, and due to the dramatic losses suffered by private, but also public financing bodies, the original public VC policies were reconsidered. In 2003 KfW was merged with DtA (Deutsche Ausgleichsbank or German Equalization Bank – TBG's mother institution). At this point, TBG seized its investment activity. KfW's own VC activities were re-shaped. Since 2004, equity capital has been delivered by KfW's 'ERP[6] Start Funds'. This scheme is exclusively co-investment-oriented and highly structured. Roughly €60 million per year are invested in young technology companies across all branches and sectors.

A second, likewise dramatic impulse was the financial crisis of 2007 and 2008 and particularly the bankruptcy of Lehmann Brothers, through which KfW suffered heavy losses.[7] Since then, KfW's leadership pressed for changes in internal organization. It did so by pursuing a transformation from a hybrid between banking and public funding administration into a formally recognized bank. As a member of KfW's venture finance team pointed out, this shift was a highly political move and not based on practical or factual necessity. Actively submitting to banking regulations such as Basel III, with all the scrutiny and limitations they impose on investment activity, was rather an effort to restore trust and to prevent scandals in the future.

> What could happen is that in the next enquiry committee, some politicians start a roar, say, 'If you had only forced KfW to submit to Basel III or the KWG,[8] then the whole Lehmann bust wouldn't have happened!' I mean there is a Lehmann trauma here. (Interview 6–4)

With the transition within KfW, decision-making processes in equity financing were altered. Investment decisions are now more distributed across different actors and evaluative practices. Some of these are perceived as adversarial to innovation-oriented financing by investment managers.

> KfW acts more like a bank [...] It is a bank after all. And if you look at what is currently being regulated in the banking sector, and under which obligations and procedural rules banks act, there is no wonder why less

comes out. There is an inertness [...] which is clearly induced by regulation. [...] There is a systemic incompatibility between start-up entrepreneurialism and the banking sector. It just doesn't fit. [...] One of the
points that are being regulated is the second vote. For larger investments,
we now need to get a second vote. There used to be only one vote on the
market side: yes, we do it or no, we don't. Now we need to ask risk
controllers in our organization. And you can hear what they are called:
'risk controllers'. Of course, they see a bandit behind every bush, these
chaps. Firstly, it takes longer and secondly, it can happen that you come
back with a 'no'. (Interview 6–4)

At the time of data collection, KfW had investments in roughly 70
companies in the field of life science. A considerable number of these
were drug-developing biotechnology companies. However, these were to
the largest part investments in late stages which KfW had entered into a
considerable time earlier. Recent years had seen very little biotechnology
activity. The reason for this is KfW's relational positioning as a 'market
follower', that is, an investor who depends on leading co-investors. The
point I wish to make in the following paragraphs is that KfW, in order
to invest in biotechnology, is dependent on a working, classic VC business model in Germany. KfW's main contributions to biotech investment,
according to a representative, are continuity in capital supply, size and
thus a positive kind of inertia in a volatile environment. However, I
would argue that in the absence of relationally fitting partners, KfW
unintentionally contributes to a narrowing of possibility spaces at the
intersection of financial and other value appraisals in biotechnology.

KfW's investment practice, as portrayed by an investment manager, is
based on financial or value-related self-interest only to a very small
degree; investment activity is expected not to produce losses overall.
There is no explicit return expectation and no limitation regarding business model or technology field. In all aspects – contract design, invested
sum, investment time period – KfW investments are meant to match the
activities of private lead investors. Within an investment consortium,
KfW stays largely passive, and, for example, does not assume a position
in an investee company's supervisory board. While representatives of
KfW are present in a wide range of communication circuits related to
start-up financing – business plan competitions, partnering conferences
etc. – KfW will only invest once a private investor is found who takes the
lead. The rationale of this approach is to add mass to existent investment
activities and thus to increase the reach and endurance of entrepreneurs
and investors.

Accordingly, due-diligence procedures are limited to financial considerations, such as the solvency, financial endowment and experience of
the lead investor or the realism of turnover expectations laid out in the
business plan. Explicitly, KfW does not perform a technology evaluation.

The point is rather that you might have an inexperienced investor in a difficult technology field. Let's stick with life sciences. If somebody comes along and says, 'I don't know, my granddad had a tarmac company, and now I wish to invest in a life science company because I have so much money', then I would say, 'Better not to touch it'. That would be a clear reason for rejection. [...] [*Or*] we consider the planning which was presented to us implausible. That would be something like €15 million turnover in year three. Aha, interesting, never seen that; doesn't happen either. (Interview 6–4)

Mirroring and somewhat contradicting this apparent openness, KfW's investment approach is highly specific, selective and normative regarding the envisioned lead investors and investment situations.

The range of possible lead investors is rather slim. Private investors and business angels tend to be too weak financially, especially since biotechnology investments usually require several, increasingly large investment rounds. According to EU competition law, each investment round can only contain a maximum of 50 per cent of public money, which makes the selection of a well-endowed lead investor crucial. For the same reason the existent public venture funds at the state (Land) level are unsuitable as partners, at least in the absence of private investors. Interestingly, *Sparkassen* count as private investors, in spite of their semi-public nature and their explicit mission to foster regional economic development. However, they too have drastically reduced their VC activities.

The large VC family offices are perceived to act alone or in collaboration with very select partners such as RETAIL VC. RETAIL VC, in turn, is unlikely as a partner for KfW due to its private household-based financing model, which is deemed inadequate. A corporate VC fund is only considered as a potential partner when it acts in the logic of a 'pure-bred' VC fund, that is, it plans to perform an exit on the capital market and does not integrate the respective investee company into the mother corporation's own organizational structure. The latter would endanger KfW's own exit. Accordingly, strategic investors like industrial corporations in general are not considered potential investment partners. The only remaining organizations are classic, life-science-oriented (and more specifically biotechnology-oriented) VC companies with headquarters or at least branch offices in Germany. As pointed out before, very few of these exist.

In addition, KfW's investment approach displays a peculiar combination of specificity and non-specificity with regard to investment and disinvestment situations. Initial investments are limited to companies younger than ten years of age and smaller than 50 employees in Germany. Founders or 'technology carriers' are supposed to hold at least 25 per cent of the company's capital. Hence, the image of the young, founder-led technology-driven start-up is invoked. KfW is supposed to disinvest after a maximum of ten years according to its own rules.

In contrast to this high specificity of the envisioned initial investment situation, KfW's investment philosophy is unspecific regarding the consequent unfolding of events. Limiting metrics such as company age, size and founder ownership do not apply in consequent financing rounds. The rule to disinvest after ten years is not enforced strictly, if situational circumstances favour a different course of action. While every investment round requires a separate decision and thus offers the opportunity to refrain from further engagement with an investee company, there are no specified valuation principles or criteria such as business model, key technologies, commercialization strategy, social and ecologic impact, jobs created, intellectual property (IP) and revenue appropriation, the actors who do it or the places where it is done. The non-specificity regarding the investment and innovation path is understood as necessary flexibility in the face of essentially unpredictable events.

> Yes, you could say, when eventually Pfizer absorbs something that was mostly developed in Germany, right, better than nothing. I wouldn't call it a conflict of aims. In Cologne they say, 'It comes as it comes'. [*'Et kütt wie et kütt'. Local Cologne dialect* [9]] You don't know in advance. At least, for ten years, [...] a few people in companies have worked, paid their mortgages and so on; fed their families. Great! (Interview 6–4)

In sum, in limiting itself to pure-bred VC funds present in the German territory as partners, KfW reproduces their selectivity and their transformative effects on biotechnology innovations. The implicit rationale of KfW investment is of course to foster innovation and economic development, although these aims are not operationalised in valuation practices other than the specifications of the initial investment decision. The implicit assumption behind KfW's investment practice is that the congruence of financial valuation and economic policy-oriented valuation is inherent in the VC business model, at least at a collective level. Therefore, it appears that there is a longing for the ideal-typical, *normal* biotechnology-oriented venture capitalist.

> The wish is that in a, I would say, heavily knowledge-based scene like biotechnology of course you need wise investors who do this in a serial fashion and not here and there a little. Here especially, we need trained and specialized investors, who invest continuously and serially. In other words, it is desirable to have the institutional VC fund who simply knows the subject matter. Like TVM in earlier days. [...] The question is: what is the normal case? And the normal case does not exist, because there are still no corresponding business models in venture capital, in life science, in pharmaceuticals, which really fit. There are exceptional attempts. Hopp and [SOUTHWEST FAMILY VC] are the most visible exceptional, individual attempts. Apart from them, there are a lot of individual experiments. (Interview 6–4)

One more specific, proactive and early-stage focused development in VC policy in the mid-2000s was the establishment of High-Tech Gründerfonds (HTGF) or High-Tech Entrepreneurs Fund. In the aftermath of the dotcom crisis, a particular area of concern was the market for seed financing. In the perception of policy makers, private seed investors – business angels and venture capitalists – were increasingly unwilling or unable to bear the high risk of early investments in technology-oriented start-up companies. The Social Democrat-led government had initiated a collaboration process named 'partners for innovation' in which, among others, government ministries and leading German industrial corporations were involved. One of the working groups established addressed the issue of seed financing and found that the best way to address the 'market failure' would be to create a national, public-private seed fund. High-Tech Gründerfonds, launched in 2005, was the result of this corporatist process of policy formulation.

The fund was equipped with roughly €270 million of capital, around 90 per cent of which were tax money. The remainder was provided by a select group of large industrial corporations such as Siemens and BASF, who became limited partners. The BMWi was to survey the fund politically. The fund was to invest in newly created German companies no longer than 12 months old and based upon technology driven R&D projects. First rounds were limited to €500.000, subsequent investment rounds to €1 million. The main rationale behind HTGF was to revive the seed capital market. A likewise crucial aim was to give the German industry advanced access to the newest technological developments: Given the agreement of company founders, the limited partners were to receive exclusive information about the deal flow (over which the limited partners have no power) and the technological characteristics of founding projects. They were envisioned as potential pioneer users, collaboration partners (build-up partnership) and strategic investors who could provide exit opportunities to the fund.

Therefore, HTGF's investment logic bears some similarity to that of a corporate VC fund. The strong position of the German corporate elite in this constellation can be interpreted in different ways. Arguing with Weber and Weber (2007) it can be understood as a way to increase the likelihood of a relational fit and thus radical innovation. On the other hand, arguing with Wu (2011), anti-disruptive tendencies can be expected, such as the 'Cronos effect': Large industrial incumbents attempt to either prevent disruptive innovations, which would threaten their business models, or gradually appropriate, absorb and mould them in such a way that they are compatible with existent business models. Given the German industry's specialization in capital goods, for example industrial production equipment for established international markets, innovations which would depend on the creation of new markets and new user or consumption practices are less likely to succeed. None of the now 16 limited partners are pharmaceutical corporations; however, two chemicals companies are among them.

In the absence of time-sensitive institutional investors HTGF's investment time horizon is substantially longer than that of a classic VC fund: An investment can be held for a maximum 12 years with an option for a two-year extension. Nevertheless, a representative stressed the congruence of interests between HTGF and regular venture funds, firstly by pointing out that the investment duration ideally is shorter and secondly by highlighting the fund's goal to be profitable. Furthermore, every single investee company is required to work as an independent financial asset. Hence, as in other examples of public VC, HTGF attempts to create relational proximity to regular VC by pursuing a similar notion of value. Its politically ascribed role is to act as an intermediary between the area of 'high risk' seed financing and conventional VC. The fund invests in start-up companies and then attempts to attract further investors. As in other cases of public VC a fundamental overlap between the aims of profitable investment, innovation and territorial economic development is implicitly presupposed and 'technology' is perceived as the hinge linking these orders. Yet, despite the efforts in creating proximity with private VC, the representative described differences:

> Our return expectation is certainly lower in the sense that we understand ourselves more as a technology fund than a classic venture capital fund. This means that we are ready to address niche markets, too, that it doesn't have to be a blockbuster market, that in our expectation, when we look at a deal, a company doesn't have to make a business case right away in which a tenfold multiple is prognosticated. Instead, we are ready to support a solid business in a small but stable market. (Interview 8–2)

In 2011 a second fund was set up with more industrial involvement (both in terms of money and number of partners) and the first fund gradually began disinvesting around 2013. Hence, HTGF although being organized in temporary funds, has a long-term perspective. A 2010 evaluation found that many goals of the fund were successfully met (Geyer and Heimer 2010). It stated, however, that the fund was less successful in supporting biotechnology entrepreneurialism. This was largely attributed to the formal restrictions on financing volume and time of first investment. Biotechnology investments are handled by HTGF's 'life science, nanotechnology, chemicals and materials' team, i.e. a rather eclectic composition of technology fields with internal relatedness. Roughly a quarter are biotechnology companies pursuing various business models.

HTGF enacts and recreates territoriality in manifold ways, the first of which is its impact on the resurrection of a national seed capital market. HTGF quantitatively dominates seed financing in Germany through its sheer size. Surprisingly the evaluation found no crowding out effects (Geyer and Heimer 2010). The reason may be found in HTGF's intensive relational work to build a seed market, to create investment opportunities (see Wray 2012) and to enrol a wide range of actors as market players. This is done by brokerage. HTGF sustains a network of 'coaches', of experienced founders and business angels from which founders can choose a fitting, designated coach. It is also done by curating in a national context: HTGF holds 'family days' on which private investors and business angels are presented selected, 'fitting' investee companies. It is furthermore done by convincing: HTGF representatives work intensively with German universities to approach scientists in with an interest in application and turn them into founders. It is finally done by co-investing with private seed investors in first rounds.

These activities were found effective in stimulating seed investment, but at the same time the specific mode of doing so was also subject to heavy criticism, both in the evaluation and my own interviews. HTGF requires founders to co-invest and also enrols friends and families of founders as private investors to create a financing round. It applies a standardized contractual design through which it protects itself (and following investors who copy the design) against dilution, but not the founder. In sum, a high degree of risk and liability is imposed on founders' personal lives; an instance which is in line with the often-described moralization and stigmatization of entrepreneurial failure in German society.

In selecting and evaluating potential investee companies, an emphasis is placed on technology evaluation. This is done predominantly with the help of external experts. Since many technological approaches funded by HTGF are based on scientific novelties with no practical validation, a profound scientific assessment has a high priority. HTGF managers thus engage with the epistemic communities, in which the respective approach is embedded, carefully balancing the views of different experts. Crucially, neither colleagues who are very close ('buddies') to the respective entrepreneur, nor rivals in the discipline come into consideration as examiners. While epistemic communities in general are transnational, two factors lend a territorial aspect to the relational work of evaluating investment proposals. Firstly, HTGF has particularly strong ties to German universities. Secondly, in identifying and approaching scientific and industrial experts, HTGF makes extensive use of intermediaries. These intermediaries are national institutions and their formalized networks, such as Steinbeis Foundation, a technology transfer organisation.

HTGF is profit oriented and not obligated to provide financing in each case. However, the fund is subject to political legitimization processes, setting it apart from private funds. In the course of the 2010 evaluation (ibid.), both entrepreneurs who received financing and entrepreneurs rejected were asked about their experiences. Issues of fairness and transparency were raised. Thus, although there is no such thing as eligibility to funding by HTGF, questions of distributive justice and the rules underpinning decision making are part of the political discussion which is crucial to the fund's legitimacy.

In its investment strategies, HTGF attempts to build larger investment consortia following its initial investment. While first investments are usually done in collaboration with private investors (business angels), other public investment bodies, the respective founder as well as friends and family, in later rounds classic venture capitalists are envisioned as partners. In such consequent rounds, HTGF remains enrolled as a minority equity holder, but 'hands over the reins'. The typical exit is a sale to a large corporation ('trade sale'). HTGF markets its portfolio both to co-investors and to strategic partners. Although the perspective is international, the bulk of the associated relational work takes place in Germany. On marketing and networking events held by HTGF, international partners are present, but typically those who have affiliations in Germany. According to a representative, HTGF's VC collaborations are heavily concentrated in Germany and central Europe.

In the field of biotechnology, both aspects, co-investments and exit strategies, were found to function in an unsatisfactory manner in the 2010 evaluation (ibid.). The absence of relevant international strategic partners for biotechnology companies in the fund's relationship network was found to be a factor limiting its non-monetary value to investee firms. Of the biotechnology firms in HTGF's portfolio, the large majority were co-financed exclusively by German public and private VC funds.

As for many other important developments, the intermediate period between the financial crises of 2000–2001 and 2008–2009 was a crucial one for public VC policy in Germany. Responding to a perceived lack of equity capital, both federal (Bund) and state (Land) governments initiated new VC initiatives. Federalism as a territorial ordering principle features strongly in the spatial organization of Germany's capital market (Wójcik 2002). The states' 'Landesbanken' act as financial service providers to local communities and state governments. Since the early 2000s, increasingly, state development banks were instituted as independent financing bodies to foster regional economic development. These development banks, like Landesbanken or KfW, are public institutions ('Anstalten des öffentlichen Rechts'). They enjoy the advantages of public institutions, such as favourable refinancing conditions, and in turn, limit their activities to support activities for the regional business landscape in accordance with EU competition law. Examples for such bodies are Investitionsbank Berlin (IBB) and North-Rhine-Westphalia (NRW) Bank.

State development banks began setting up VC funds in the mid-2000s. These funds were to provide equity capital to companies within the territorial borders of the state. At the same time, their purpose is to act as local lead investors and to help enrol further investors so as to create larger, more effective consortia. These investment partners are not formally limited to the state territory and can, theoretically, be based anywhere. It is noteworthy that, in isolated cases, state development banks also invest in private German VC funds as a 'softer' measure of VC policy. These VC funds than do not operate under a formal territorial constraint, but under an implied stakeholder interest in fostering investment in particular regions. One example of this practice is a Berlin-based VC fund with a focus on platform technologies, diagnostics and medical technologies, among whose investors are the development banks of Berlin and North Rhine-Westphalia, and whose investments are almost exclusively located in NRW and Berlin/Brandenburg. In sum, the rationale behind state VC funds is not to demonstrate the existence of investment opportunities as in the case of British RVCFs (Sunley et al. 2005), but to meet a demand which is not satisfied by 'the market', and to mobilize additional supply.

Typically, these state funds do not address seed financing, for two reasons. Firstly, in the same period (around 2005), the federal government established a national technology-oriented seed fund called High-Tech Gründerfonds (High-Tech Entrepreneurs Fund). Secondly, public and semi-public seed financing is often organized more locally. A diverse and institutionally heterogeneous landscape of seed financing has emerged involving business angels, seed funds rooted in cluster initiatives (like BioM in Munich), public and semi-public banks. This is not to say that seed financing is available everywhere. Instead, it appears to be rather concentrated in the centres of economic activity.

In the state of North Rhine-Westphalia, for instance, eight local seed funds with funding from the state development bank (NRW Bank) were established. They were located purposefully in the state's urban and economic hotspots, following perceived demand. These seed funds are managed independently of NRW Bank by local investment managers: individuals from various backgrounds (e.g. *Sparkassen*) whose decisive quality is that they have made a name for themselves in equity investment and that they are well networked. Such activities can be viewed in the context of 'place shaping' (see Chapter 3). State VC funds make a point of not automatically investing in companies which have previously been funded by public seed funds, but to subject them to the same, non-discriminatory evaluation procedures as every other potential investee firm. Nevertheless, this form of progression is considered normal.

Regional public VC funds have been a topic of discussion for a considerable amount of time. Important questions are whether a territorial investment focus is productive in fostering entrepreneurial activity (which can benefit from exchange across regional boundaries, Wray 2012), and whether RVCFs attract or rather discourage private investments ('crowding out').

The perspective of fund representatives is that by offering a local lead and intimate knowledge of investee firms and the regional economy, they act as enablers of private investment. The latter bring both funds and experience, often across physical distance, and thus can establish relational proximity to potential customers and markets. Like corporate venture funds, public VCs mimic classic VC investors in order to demonstrate proximity, win trust and appear as relationally fitting partners. Representatives interviewed stressed that their investment activity was profit-oriented 'like every other investor' and hence their interests were compatible with those of co-investing private funds. They also conceded that, as public bodies, they had special responsibilities to the region, usually accepted slightly lower returns than private investors, were more patient in their investment time horizons and invested earlier than private VC funds, in situations of perceived 'higher risk'. When setting themselves apart from classic venture capitalists, public VC managers repeatedly employed the figure of the 'hockey stick': a value development curve which slopes only slightly in the beginning and later turns to exponential growth. Expectations of this type of growth were attributed to classic venture capitalists, while own growth expectations were described as more modest. Some public VC funds have limited durations while others are open-ended 'evergreen funds'.

Public VC funds were subject to criticism by several respondents. One point of attack was the perceived lack of entrepreneurial experience on the side of public investment managers (indeed the three Land- and one Bund-level public VC investment managers interviewed by me did not have an entrepreneurial track record), and the fact that their funds, as products of political considerations, do not really depend on commercial success. Hence, despite all efforts in mimicry, public VC is not seen as the 'real thing'. One VC investment manager claimed that there was 'already too much public money in the market', while conceding that there were very few alternatives to it. However, from the side of entrepreneurs there was also praise for public VC offices with regard to their knowledge of the respective subject matter, their financial strength and range and quality of networks.

As in the case of KfW the underlying assumption of public regional VC investment is that the value of regional economic development can be brought in congruence with the value of financial investment returns.

> We are a business development bank and therefore there is the aspect of structural policy. But we approach the issue in the conviction that we will only be successful in terms of structural policy if the companies are commercially successful. And this means that in terms of selection criteria we act like a private investor would. (Interview 6–5)

For state VC funds, the respective region's technological potentials and capabilities are the starting point. In their logic, regional economic development can be fostered by commercializing regional knowledge and by drawing on existing regional strengths. IBB's VC office, for example, is structured into teams which each represent a field of regional specialization as outlined in the Land's cluster strategy.

One of these is life sciences. Generally, life sciences, including biotechnology but also medical technology, feature prominently in state VC activities. While Land VC funds do invest in biopharmaceutical product development, several respondents expressed concerns regarding the cost, risk and duration of such investments. NRW Bank's VC facility, for instance, has a total of €120 million under management, which is large compared to existing private VC funds, but small compared to average fund sizes shortly before the dotcom crisis and also small compared to the costs of biopharmaceutical product development. The need to build strong investor consortia and to identify development paths with reduced regulatory complexity (such as orphan drugs, see Chapter 4) was stressed.

Since public regional VC funds are bound in their investment decisions by territorial borders, they are required to identify all investment opportunities within this territory. As a consequence, even a dedicated life science team is likely to be less specialized than a private, non-territorially bounded life science VC office. At the same time, factual and potential inner-regional technology interdependencies are naturally subject to investment related considerations. Within an investment portfolio, each investment is required to stand alone as a financial asset. Yet, experiences with investee firms, contacts, networks and actors present within the territory frequently serve as reference points in evaluations of new investee companies. In the cases studied, long relational histories were relevant for investments in biotechnology.

Land VC fund investment managers differed in the ways in which they described evaluation procedures of potential investee firms: technology, business model and market potentials were either subject to highly structured evaluative procedures (expert interviews, online research), or were viewed in a more holistic and intuition-based perspective. Likewise, the question of which aspects of due diligence were performed in-house was answered differently. Not all investment managers were trained as scientists.

Land-level VC funds act as lead investors with active roles in companies' strategic development. Typically, their representatives assume positions in supervisory boards. While they do not favour and actively pursue particular business models and commercialization strategies, their influence on these particular innovation processes can be characterized as a cautionary, conservative yet company-building-oriented one. Strategic investments as well as revenues from service and licensing deals are seen as the preeminent sources of a profitable exit.

Notes

1 Referring to the type of shoe and thus indicating a rather casual attire.
2 Literally 'medium sized company'. However, the term carries culturally embedded connotations. 'Mittelstand' stands for a traditional and highly regarded understanding of entrepreneurship enacted by mostly family-owned, long-lasting and long-term oriented, highly competitive manufacturing companies predominantly utilizing vocational skills and training schemes.
3 Cooperation with regional banks, private as well as semi-public ('*Sparkassen*'), was described as a frequently used vehicle for regional investment as well as philanthropism by wealthy individuals by one respondent.
4 In the original German, 'Tante Emma' is the prototypical small corner shop owner, and in this context serves as a metaphor for the small investor in need of protection. In the aftermath of the Lehmann bankruptcy, female seniors with small pensions, who had been talked into investing in allegedly safe Lehmann products by their local Sparkasse agents, received a high level of media coverage.
5 See next sub-section.
6 ERP stands for European Recovery Programme or 'Marshall Plan'. The funds stem from the original Marshall Plan estate.
7 Shortly before Lehmann's fall in 2007, KfW had transferred €319 million to the US bank as part of a currency transaction, although allegedly KfW had been aware of Lehmann's situation. Consequently, KfW became the centre of a scandal.
8 Kreditwesengesetz or Credit System Act.
9 KfW is based in Bonn, near Cologne.

3 Relational dynamics in early stages of biotech innovations

With this chapter I turn to the individual case studies, particularly their early stages. I would like to make the following arguments: In the early stages of biotech innovation, future technology assemblages – new technological solutions 'living' in multiple interconnected places and contexts – are conceived. They come into existence as actor-networks of people, artefacts, practices (often new ones), organic matter and organizational arrangements, which, for the first time, enact a new technology. These actor-networks are *local* in at least two ways. Their reach and validity remain limited to a preliminary, experimental stage. Their existence hinges on being embedded in a specific context, which may include more than one place, but which is specific to geographical location(s). Furthermore, the emergence of situations in which innovations can be conceived, the gradual overlapping and interpenetration of social spaces in which the important 'conversations' (Rutten 2017) can take place, is tied to the evolution of organizational and institutional settings which only unfolds in specific places: Locales of overlap between different practices and institutional orders ('heterarchy', Stark 2009), locations of technological and institutional resources. If a relational approach is to be taken seriously, this means that the content, the inner structure of an innovation is sensitive to these local-specific dynamics of emergence. In practical terms, the 'early stages' discussed here encompass the creation of new technological areas and opportunities, the creation of a biotechnology start-up attempting to capitalize one such opportunity as well as attempts to stabilize the pursued approach in the immediate aftermath. This phase ends with the advent of a transformation, often involving a severe crisis, from a technology-driven venture to a truly business-centric operation.

I will tell the story of unfolding events by focusing on the dynamic interplay, the tension and fertilization between two modes of agency, two logics of appreciating and creating value, both performed by key individuals: Firstly, a dynamism of creating and maintaining open spaces which allow the emergence of new ideas and, more importantly, generate opportunities for idea utilization beyond conventional institutional and organizational repertoires of action. I will refer to this dynamic of agency as 'place shaping'. It typically involves people higher up in the hierarchies of science, industry and business promotion, who cultivate cutting-edge research as well as high levels of acceptance for 'misfits' within their area of influence. And secondly, a dynamism of recognizing and exploiting opportunities in specific times and places for one's personal economic, reputational, but also ideational and civic goals. Typically, this opportunity seeking and exploitation is driven by an ambition to deliver a transformative new approach, a technology which will change entire fields of application. I will thus refer to this modus of agency as 'enthusiasm-driven entrepreneurship'. It typically involves a company founder who brings together a problem ('obstacle problem', see 1.4) and a prospective solution, develops a vision, wins supporters and builds a company. This narration represents a more stylized attempt at characterizing distributed roles and interactions in innovation processes than Van de Ven et al.'s (1999) study of innovation processes. Nevertheless, there are overlaps.

Investors and investment are part of this interplay. Investment in early stages of innovation requires more than 'financial' interests and approaches of valuation. It requires forms of intrinsic motivation in conjunction with business interest. This is not new. Indeed, the often heard 'family, friends and fools' represents a category of intrinsically motivated or plainly misguided investors who, seemingly without the proper professional distance, invest in extremely risky ventures ahead of more formal venture capital (VC). What I wish to show, however, is that investors and investment value appraisals are just as diverse, as dependent on local and temporal context as the technologies and ventures themselves. Enrolling investors is part of the general efforts of 'translation' – joining actors together under a common mission by wrestling them from competing ventures – which constitute disruptive enthusiasm. Thus, investors are both actors and objects of action in a highly context-sensitive setting.

To highlight the dependence of an innovation's emergence on long-term developments of organizational and institutional structures, but also the remarkable scope for individual agency, I will begin the presentation of each case by characterizing the institutional environment of idea emergence. In some cases, this requires going back in time even further than the earliest cases and considering the early attempts of German governments to create a biotechnology industry virtually out of nothing.

3.1 Max Planck Institutes and Gene Centres as Nuclei

In the early 1980s three 'Gene Centres' were set up in West Berlin, Heidelberg and Munich. These centres represent an early attempt to enrol industrial corporations for biotechnological research investment. They were envisioned as first-rate, well-funded research facilities, which were to create a push to innovation dynamics. The centres were half publicly funded and half funded by the industry. Although the expected impact did not directly materialize, the centres would later become important nuclei. In the early 1990s, Germany was in the midst of its post-unification recession. A lagging behind in high-tech industries was universally felt. In addition, industrial corporations had begun engaging with biotechnology – by building subsidiaries or investing in start-ups in the United States. Corporations justified this move with the uncertain, partially adversarial and fragmented regulatory situation in Germany (Adleberger 1999).

In this situation, the Federal Ministry for Science and Education (BMBF) initiated the BioRegio competition. In this programme, regional initiatives were financially and administratively supported in the development of regional networking and consulting activities to initiate biotech entrepreneurialism and to formulate an appropriate strategy. Crucially, these initiatives had to focus on small-scale, more functional interaction spaces and not *Länder* territories. All participants were asked to submit their regional strategy proposals. The three best ones were to receive DM 50 million in support funds – which had to be co-financed – to further develop their respective clusters or networks. The prospect of such substantial funding put pressure on regional policy makers to support biotechnology (Adleberger 1999; Dohse 2000).

The regions of Munich, the Rhineland and Heidelberg emerged victorious from the competition. These regions had been comparatively strong before (two possessed Gene Centres) and were further strengthened through the allocated funds. More importantly, however, in many German regions, serious political and economic support for collaborative action across institutional boundaries had been mobilized. The BioRegio competition is therefore credited with being the key to the surge in biotechnological company founding and investment in Germany in the late 1990s (Dohse 2000; Müller 2002). Both Gene Centres and the BioRegio competition feature strongly in several cases. Specifically, the institutional history of BioRegio and Gene Centres opened scope for agency by key individuals (Gailing and Ibert 2016) in creating support environments for innovation, including business incubation and early stage investment.

In case 1 ENCAPSULATION, the interplay between place shaping and enthusiasm-led entrepreneurship unfolded in the following way: A professor of biology, henceforth referred to as the 'place shaper', had frequent contact with industrial research and development (R&D) while working at the local Gene Centre, which was affiliated with Munich's highly reputed Ludwigs Maximilian University (LMU). He, too, became a company founder and, in the early 1990s, set up one of the first German biotechnology companies. Incidentally, this company became the first of a growing number of biotech start-ups in the village of Martinsried outside Munich, which, with its business incubator and multiple research institutes, today is the heart of Munich's biotechnology cluster.

> The reason why this centre developed here in Martinsried essentially is a consequence of us founding [*company name*]. I was a professor at Gene Centre. And regarding physical activity, I am a lazy person. I said I need the company in walking distance. (Interview 1–3)

During the BioRegio competition of 1994, the founder participated in developing the Munich biotech cluster strategy, which won the competition and brought substantial funds to the city. Munich is Germany's preeminent high-tech cluster. The involvement of – locally available – high-level industry representatives in strategy formulation led to a more daring approach in cluster support policy compared to other German regions. Substantial funds, which were gained in the competition, were co-funded by the industrial sector to create a seed fund for biotechnology companies. This seed fund's investment board brought together representatives from the local pharmaceutical and chemicals industry, as well as the local VC sector, which is likewise comparatively strong in Munich. Entrepreneurial risk-taking by scientists would from now on be valued and evaluated in a very hands-on fashion. The cluster agency created during the BioRegio competition combined networking activities with a business incubator and said seed fund. The aforementioned 'place shaper' became the agency's as well as the seed fund's leader.

Meanwhile in Berlin, a physicist who had no previous experience with the pharmaceutical industry, was working in the R&D department of a local pharmaceutical company. He was charged with creating an Encapsulation for a new active ingredient.[1] While trying to follow instructions, he realized that, from his perspective as a physicist, he could not accept the approach his colleagues and superiors were pursuing.

> You have to consider, I was in pharmaceutical technology. This pharmaceutical technology in principle has several fields, and they are all pharmacists. [...] And they told me to develop a formulation for [a product]. And then they told me this and this had to be done. And I reproduced, reproduced, and I couldn't make it work. I was at a point where I turned everything towards Mekka, but that didn't work either. Eventually I said; 'I refuse to do this, I won't do this anymore, this is shit, this is frustrating, it doesn't work. The physics is wrong'. (Interview 1–1)

To delve deeper into the problem he took time off. Reading and experimenting alone he realized that the substance he was working with had other properties than the physics books claimed. This discovery would allow for a radically improved Encapsulation procedure.

> And then the head of the department said, 'You physicists are such snobs, go and make it better!' So I said, 'Okay'. So he gave me three months leave. I went to [...] the TU library and other university libraries. I read. I had discussions with a lot of people. At some point I had the idea, this could work. All physics books say that this is impossible, but I said, That's okay, we can try it. It really worked. (Interview 1–1)

In essence, his discovery was that a particular Encapsulation polymer had radically different dissolution characteristics from what physics textbooks predicted (a knowledge domain that was being technically applied, but not critically interrogated by pharmacists). The physicist described this discovery as both stunningly simple and very versatile in terms of application, which could include, for example, the Encapsulation of active ingredients in washing powder. He had a strong preference for exploring and exploiting the discovery outside his employing company, and found an opportunity to do so when he was approached by a team of founders, some of whom had made careers in the pharmaceutical industry, some of whom were linked by kinship. They hired him as a researcher and set up a company in Munich to exploit the technology. The intellectual property (IP) was, favourable, handed over by the original employer. Munich, in contrast to Berlin, possessed business incubation structures at the time. The start-up received an initial investment from the local biotechnology seed fund. In addition, the support agency drew down funding from a national early stage investment program which required co-funding. The place shaper explained the decision to fund the company:

> We as BioM were positioned in such a way that we were really open for new things. Und we wanted to take risks. Of course, there were doubts, but we really liked this [*technology*]. If this really works, it will be fantastic. And we are proud of the fact that out of the 40 investments we had here, 12 to 15 went into insolvency – but not because of their technologies. [...] There was [...] a risk, because we were no formulation experts. We only tried to comprehend this with chemical expertise. We said, 'Yes, this could work, but the other guy is the expert'. There are not many experts in this field. And when we had a positive vote from the pharmaceutical industry, we went ahead and did it. (Interview 1–3)

The local environment and particularly the mentorship of the place shaper became highly influential for the further unfolding of the innovation. It helped free the idea and its carrier from inhibiting environments, helped mobilize both and provided both guiding and stimulating opportunities for interaction. Under the start-up's original leadership, the physicist soon felt overly restricted, deprived of his freedom of scope, exploited and positively duped. Various respondents confirmed that this perception was not wrong. In the place shaper, he found an ally and a mentor. After being urged by the physicist, the place shaper, in his formal executive position, agreed to buy out the original leaders (including IP) and put the physicist in charge. He explained his stance by invoking their shared background in science as well as his experience that bad management often poses a bigger risk to companies than faulty technology:

> If I hadn't been a scientist before that and hadn't been able to identify with him on a cognitive level, it might have gone differently, because usually you stick with the managing director. On the other hand, we were intelligent enough to say, 'The company lives and dies with Mr [*name*]. If

he quits here and says, "This is it, I quit and go back to [*company name*]", then this is the end'. (Interview 1–3)

After being liberated in this way, the physicist embarked on a search and discovery process to put his idea into practice. He began to travel, sought out old associates and sponsors, including his PhD supervisor in Aachen, and engaged in heated debates about his plan. He also used these contacts to assemble a young, small team, which included a specialist in bio-physical chemistry and an engineer with knowledge of the technicalities of making solutions. They came to Munich, attracted by independence, the physicist's charisma and a bold vision. Respondents provided vivid narratives of how the local environment provided inspiration and opportunities to make contacts, but also of the physicist's aptness in making use of this environment.

> Here in this small space, in this backwater, Matrinsried, roughly 70 to 80 companies are located. […] There is a density of course. You get to meet people. And Mr [*name*] is a communicative person of course. He goes for dinner and talks to this one and that one. Then he walks over to the Max Planck Institute, finds the peptide chemists. This is a huge advantage. […] He can convince people. Well, he talks you to pieces. Either you accept it or it gets on your nerves. (Interview 1–3)

The strong presence of biomedical research, biotechnology companies and pharmaceutical industry in the Martinsried cluster geared the technology towards applications in the biopharmaceutical sector, although other applications would have been possible. The practical difficulties of bringing one of the new, complex organic molecules developed by biotechnology start-ups as future medical drugs, intrigued the physicist. He relished engaging not only in the sharing of knowledge, but also in experiences of trial-and-error, of setbacks, and failure that were locally accessible to him.

> At the beginning we told ourselves, there are many biotechnology companies. I mean, here on the campus are so many; they have such wonderful proteins, but these are instable. They are stable at minus 20 degrees Celsius or whatever. If you want to test something in humans later, it has to be in a way that they feel good with it. There has to be a viable form [*of delivery*], a pill, a capsule or an injection. This is what we create. (Interview 1–1)

The combination of mobility and local contacts – which included local 'temporary buzz' (Bathelt and Schuldt 2010) with mobile individuals from faraway places – enabled the first contacts to potential customers and investors beyond the seed stage. In 2001, a California-based pharmaceutical company with a specialization in inhalable drugs invested in the physicist's company in Munich, to have a new formulation for an active ingredient developed. A new formulation meant that the drug could be inhaled in longer intervals, with fewer discomfort or fewer side effects or all the named together. Thus, an idea with a high degree of scientific newness was to be used to improve an existing product with an existing active ingredient, but with noticeable practical effect. This logic was encountered repeatedly in the research for this book: Corporations sometimes invest in radically new technologies following an anti-disruptive rationale, seeking to augment existing technology and aiming to maintain an existing market position.

With the new investor, who was located comfortably far away, and with a collaboration partner inside the investing company who was described as a 'buddy', the physicist was able to continue his pursuit in a fairly autonomous manner.

Many of the dynamics observed in this case repeat themselves in other cases, albeit with nuanced differences. Several cases are strongly influenced by an institutional environment which characterizes the German biotechnology situation as much as BioRegio and Gene Centres: Max Planck Society. Max Planck Institutes are publicly funded research centres with a mission of advancing cutting edge basic research. Often the names of institutes (such as 'Max Planck Institute for Molecular Genetics') stand for novel epistemic approaches and emerging fields of enquiry. Setting up a Max Planck Institute is typically among the first institutionalizations of a new scientific field, and thus these institutes become nuclei for the creation of new epistemic communities at the intersection of formerly disparate ones. A Max Planck institute is centred on a strong leading figure, an outstanding and pioneering researcher with a far-reaching international reputation. Once set up, such institutes grant the respective leader an enormous degree of autonomy over the utilization of funds as well as research strategy.

However, Max Planck Society struggles with commercialization. The organization's mission is basic research in the purest sense, although since the mid-2000s efforts have been made to include technology transfer into the organization's mission. Surprisingly, Max Planck Institutes are the origins of a very large number of science-driven start-ups. The often-cited reason is that for Max Planck researchers, the only way to commercialize an idea is to leave the institution. Furthermore, the organizational format of having autonomous teams of researchers with strong leaders in the centre increases the likelihood of entire teams, following an entrepreneurially minded leader, founding a company to pursue the commercialization of an idea. Industry collaboration, on the other hand, is alien to Max Planck Society's self-perception.

This predilection for technology transfer via start-up founding is rooted in the organizations institutional history, as a Max Planck Society representative reported (Interview 5–5). Based at Max Planck-Gesellschaft's (MPG) headquarters in Munich, the sub-organization Max Planck Innovation manages all activities relevant for transfer. Crucially, it centrally manages all IP generated by Max Planck scientists. This sub-organization also runs a small-scale seed fund designed to support start-ups by Max Planck scientists. In this fund, two activities are combined: managing MPG's financial assets and supporting technology transfer. This organizational design was initiated when a Max Planck professor co-founded a company with American scientists in the United States in the early 1990s and demanded support from his mother institution. MPG's leadership respondent to this demand by allocating financial assets, which required management in any case, to a seed fund. Like BioM's seed fund, rather than achieving growth, the fund's primary goal is technology transfer via business incubation – preferably in Germany, but theoretically anywhere in the world. It is expected not to make real losses but has no profit aim beyond that. The fund typically invests small sums well below €100,000. Probably more importantly than the money invested, Max Planck Innovation in its investor role attempts to protect the interests of founding scientists vis-à-vis other investors. This approach to transfer – centralized control over IP, mentoring and small-scale investment for start-ups – reflected in several cases. The Max Planck institutional framework provides a variety of resources, but also resistance, to the agency of place shaping.

In case 5 AUTOIMMUNE, the agency of 'place shaping' was distributed across two interacting key individuals, a Max Planck Institute leader and an industrial executive. The personal relationship between the two enabled the creation of an unprecedented form of boundary spanning cooperation, which would, as an unintended side-effect, create an opportunity space for the innovation process studied in this particular case. The Max Planck researcher had pioneered the field of X-ray structural analysis over several decades, winning a Nobel Prize in the process. He had observed the proliferation of the technology in various industry sectors throughout his career and participated in formal and informal exchanges since the 1970s. The relative ease with which industry representatives could be met in Munich had been a supporting factor in building these relations. Appreciating industry involvement in research, he frowned upon the recent tendency of large pharmaceutical corporations to axe their research capacities.

His partner on the industry side was a serial boundary spanner between academia, industry and start-up entrepreneurialism. Since his days as a student of biology in Munich, the latter had been thrilled by the development of gene technology and its potential for practical application. He perceived himself as 'the practical type'. At the same time, he had found it attractive – and easy – to share in the working practices of pharmaceutical corporations, one of which ran a plant nearby. Fellow students obtained research positions in the company and shared their experiences. After starting his career in the industry and pursuing it with some success, he began to experience the bureaucratic organization of work in a large company as unsatisfying. After several years, he started his own company. Over the following years, he participated in the founding of three companies. Later, he would return to the pharmaceutical industry and advance to the position of head of research and development of a Swiss diagnostics company. In this position, he approached the Max Planck leader in the early 2000s to suggest an arrangement in which his company would fund a mixed research group on-site at the Max Planck Institute, which possessed the most advanced equipment. Its proposed purpose was to study a set of molecular binding mechanisms in which the company was interested. It would be the first such arrangement in Germany. Both participants considered it an elegant and efficient solution, given cost pressures and mutual interests.

Theoretically, both participants were separated by a substantial 'institutional distance'. Indeed, the legal departments of both organizations objected on the grounds of internal rules governing publication and patenting practices. Max Planck Society's approach to controlling IP presented an obstacle, while the Institute leader in question did not find it objectionable that an industrial corporation would acquire IP from a joint project. Both individuals perceived a common interest, were personally acquainted and, living in Munich, found it easy to negotiate the terms informally. Furthermore, they had substantial influence within their respective organizations to enforce the idea.

> And that is why I simply sat down with professor [*name*] one weekend and we said, 'Now we have been negotiating for months, especially our legal departments. But we should do it. And we will simply do it'. Then we both told our legal departments, 'It has to be done now and in this way'. (Interview 5–1)

The arrangement was giving the institute leader additional slack to hire scientists and do research freely, an important factor for the innovation which would later unfold. In 2005, an immunologist from a German university came to Munich to make use of the Max Planck Institute's capability in structural biology, specifically X-ray structural analysis, a technique to explore and visualize organic molecules' structures. This technology requires an assemblage consisting of a focused, high-powered radiation source which 'fires' at the material in question, a reliable source of the latter, either synthetic or natural, a sensitive detector to measure reflected radiation and computing devices to analyse and visualize the data. Crucially a conjunction of all these elements was locally available since the institute had been a pioneering site for this practice over several decades.

The immunologist's interest was in exploring a specific molecular receptor which is relevant in autoimmune disorders. He received a position which was funded via the collaboration scheme discussed above and the extra slack it afforded. Although the work was being funded via the unique collaboration project, the knowledge gained did not directly contribute to the funding diagnostics company's R&D agenda. Instead, the institute itself patented the findings. Later the immunologist, together with a colleague, founded a company to exploit the knowledge gained on the binding characteristics of the receptor in question and matching antibodies. The company was to deliver novel therapies for autoimmune diseases like lupus and rheumatism. Since the receptor-antibody binding mechanism in question is relevant for immune regulation in general, the start-up company was pursuing an agenda with a potentially transformative effect on a broad field of possible applications. Crucially, the Max Planck Institute leader, who had accompanied various start-ups before this one, supported and mentored the venture, but did not claim part of it for himself.

> That was the end [*of basic research*]. This had to go into commercial development. And both of them, [*founder name*] and [*founder name*] were very interested in that. Besides, this was also the trigger for the founding of [*name of an earlier start-up*]. *I was, if you will, the godfather who stands behind it and says, 'This is good, do it'. I myself want nothing to do with operating business.* (Interview 5–2, emphasis added by author)

Max Planck Society itself, via Max Planck Innovation, acted as the first small-scale investor and institutional mentor (Van de Ven et al. 1999). The organization's representative then provided crucial assistance in the search for a more substantial investment. The search concentrated on investors with an intrinsic motivation to engage with a technology at a very early stage. Rather than searching globally, a potential investor was sought in relative geographical proximity, which in this case meant Germany and adjacent European countries. As a result, a corporate investment fund run by an Italian pharmaceutical company was approached and enrolled as an investor. The pharmaceutical company, at the time, had an interest in developing new drugs using the start-up company's core technology, and the fund's specific purpose was to be a search and discovery vehicle for new technologies. Representatives of the pharmaceutical company were present on-site at the Max Planck Institute and, according to one scientist, were capable of understanding and appreciating the technology on a scientific level. Thus, as in case 1 ENCAPSU-LATION, the first substantial investment after small-scale seed investments, came from a corporation with the specific intent to commercialize the respective technology. In this case, however, the aim was to develop new products rather than improving existent ones.

Cases 3 GENE FUNCTION and 4 BIOMARKER, introduced at the beginning of this book, likewise are strongly shaped by the Max Planck institutional context. In contrast to the cases discussed above, the early stages of these innovations are centred on Berlin rather than Munich. Furthermore, in contrast to case 5 AUTOIMMUNE, the origins of these innovations date back to the 1990s, as in case 1 ENCAPSULATION. Since Berlin at this time lacked a developed support infrastructure, both cases include narratives of improvisation and handicraft when unused buildings were converted into lab spaces and supporters were enrolled without established cluster structures. Despite these differences, there are marked similarities in terms of the argument put forward in this chapter: the relevance of context and interacting dynamics of agency.

In case 3 GENE FUNCTION, respondents attributed much of the innovation's path to the place shaping agency of yet another Max Planck Institute leader, in this case the founding director of a new institute focusing on plant bio-technology near Berlin. He, according to a key member of the founding team, installed an industrial orientation in the institute's working routines. Pre-viously he had worked in the Berlin Gene Centre which had been jointly funded by the state government of Berlin and a Berlin-based pharmaceutical company. Therefore, he had participated in research activities which com-bined public and industrial logics of knowledge creation. Typical aspects of industrial work, such as a formally defined division of labour and an appre-ciation of process efficiency later re-emerged at the Max Planck Institute.

> Yes, we simply analysed our processes in a systematic way and asked ourselves where we could invest to make the process faster. We intro-duced a range of improvements. And Mr [*name of institute leader*] cov-ered certain processes with technical staff. That was something extraordinary back then that, in a research institute, a technical employee was tasked with taking care of a specific process and doing this exclu-sively and very well. Usually, a technical employee is assigned to a sci-entist and they do everything a little. So he [*institute leader*] had anticipated the industrialization a little. (Interview 3–3)

Another way the industrial logic was transferred to the everyday working routines in the institute was a frequent exchange with industry representatives. On a regular basis, scientists from industrial corporations gave presentations at the institute, an experience the researchers there found enriching. This fre-quent exchange with industry representatives had a noticeable influence on the inner structure and the development path of the technology. The research team was looking for a method to observe plant cell metabolisms. On one occasion, when representatives of a German chemicals corporation visited the institute to discuss ideas, they suggested using gas chromatography. At their company, they had applied the method to observe the responses of plants to herbicides. Subsequently, gas chromatography became an integral part of the emergent technological assembly in the Max Planck institute. It was this chemicals company which would later provide the first (and only) large-scale investment. While respondents did not see a causal linkage between the two events, they acknowledged that this pre-existent linkage put the chemicals company in a good position to appreciate the technology.

When the technology was mature enough, the institute director decided that it was now time for commercializing the results. He followed a strict delineation of the purpose of a publicly funded basic research facility. Having experienced both logics in his biography and keeping regular exchange with industry representatives, he perceived a high degree of complementarity and scope for cooperation between both sides. In the innovation process studied, he judged very early on that, after an experimental demonstration of feasibility, it would be an industrial corporation's job to take it to the market. The institute would then refocus its attention to scientifically more challenging ideas.

> So then we said, 'Yes, this is actually pretty nice, but in actuality this is a routine approach [...]'. We had demonstrated the concept. [...] And actually, if we were to repeat this 100,000 times now, this is not the job of a Max Planck Institute. I would understand the job of a Max Planck Institute in such a way that we try to create innovation. Innovations require a proof of concept and maybe some degree of implementation – but not routinization and scaling. [...] This is classically the job of the industry. (Interview 3–2)

The institute leader's perception of the division of labour between public research and industry was such that he did not see any value in patenting the new technology, which was later done by a corporation. To him, knowledge commercialization was not the purpose of a Max Planck Institute purpose, and further research could be done using a free research licence. Hence, while appreciating the practical utility of a slightly more industrial approach in public research, and while maintaining relationships to industry based on a 'service ethic', the internal appreciation of knowledge was decidedly 'basic research'-oriented. He thus felt at odds with the orientation of Max Planck Society's central leadership level, which pursued a centralized patenting strategy and valued small start-up companies over industry collaborations. The search for an investor was thus focused on industry collaborations, and given the institute's focus on plant biotechnology, an application in agriculture was envisioned.

A British postdoc in biochemistry played a part in the process from the beginning. He had sought an environment where he could work at the boundaries between different epistemic communities, namely molecular genetics and biochemistry. He found it in the newly established institute and became a key figure in the research effort.

> It was exactly the topic in the border zone. There were a lot of molecular biologists, who were trained in such a way that they did not, I would say, understand the biochemistry properly; bio-chemistry in the sense of metabolism and dynamic processes. And on the other side, the people who understood this part well typically did not have a background in molecular biology in those days. So, during the time we are discussing here, the early and mid-90s, the connection between those two areas had enormous innovation potential. [...] I think both sides sought each other. Where people analysed metabolism, they always looked for test systems. Typically, these were the ones where genetic alterations were executed. And on the other side, the colleagues who had the ability to clone genes and create genetic changes, they always searched: What impact did this

have on the biology of the system? What did I learn about this gene? And they then sought contact with physiologists, biochemists, cell biologists who could evaluate the results better. [...] I think it was the possibility to actually do something that limited us, the finance, the funds for projects and positions. [...] Through the foundation of the Max Planck Institute, we had scope for design. Then it was simply possible to buy the machines we needed for [technology] without having to go through three committees with the proposal. (Interview 3–3)

Furthermore, he had a strong interest in contributing to practical problem solutions, a desire he could not satisfy in other academic institutions. When the question of commercialization came up, he quickly decided to join the founding team, while the institute leader remained in his mentoring and supporting position.

While case 4 BIOMARKER is different in terms of the unfolding innovation path (e.g. here industrial investment, there VC), there are similar dynamics of agency. In case 4, too, a leading researcher, in this case a Max Planck department leader, acts as 'place shaper'. He had no history of industrial work, which makes him the exception in the cases studies here. Throughout his biography, he had contributed to the pioneering of new approaches in genetics, particularly epigenetics, that is, the study of gene activation and de-activation. His career path kept him at the forefront of research, but also brought him in opposition to more orthodox representatives of epistemic communities, namely conservative geneticists, who viewed the function of genes in a more deterministic way. He lobbied the German Research Foundation (DFG) to counter Germany's perceived backwardness regarding the influence of deterministic genetics vis-à-vis epigenetic approaches.

In 2001, [we] had the first DFG priority program in Germany, or actually worldwide, focusing on epigenetics [...] lasting from 2002 to 2009, a nationally coordinated program. There, we brought all working groups together who did something in this general direction. [...] Regrettably [*there is*] too little [*epigenetic research*] in Germany, there is not much done in this direction. I think Germany built certain other traditions. (Interview 4–6)

When his PhD student, who had worked on a specific measurement and analysis problem, announced that he would like to patent and commercialize this knowledge and set up a company, the place shaper supported the proposal and mentored the commercialization effort, which consisted of patenting the technology and writing up a business plan. This plan was centred on technology and multiple possible applications rather than one business case. Like other examples of place shapers, he did not want an active part, credit or revenue in the commercialization venture.

With me, he [*the founder*] found a situation where I do not feel the need to show off my ego. And I think this is not always the case in other Max Planck Institutes, because the Max Planck Society is built according to the 'king's principle'. There is a director who sets a course of action, and since you need strong personalities for that, you get strong personalities. And these strong personalities are not always beneficial to spin-off creation. Creating a spin-off means that you have to let something go. And I did not find it hard to let this go, because my heart did not cling to founding a company [*myself*]. (Interview 4–6)

The PhD student came from an entrepreneurial family. His father was an entrepreneur and he previously had participated in another biotechnology start-up together with his brother. He was described by his colleagues as extremely ambitious, charismatic and convincing in face-to-face interactions. He matched the common image of the disruptive entrepreneur. Both at his working location and via temporary co-presence in other places, such as a conference in New York, he approached potential collaborators and assembled a founding team consisting mainly of other PhDs of his age who were attracted by his charisma and shared his vision. The mentoring department leader urged the founders to take advantage of the location (both the institute and Berlin in general) to gather as many contacts and inspirations as possible. He had an indirect interest in the venture. Like other place shaping individuals, he saw engagements with application as a way to open up his own institutional space, to gain inspiration while at the same time stressing and reinforcing the boundary between basic research and application.

For me the objective was never cancer research. Now, during the last years it has grown a little, but I still use it as a tool to develop new things. I don't stand there and say, 'I will test this through until I find some Biomarker I can commercialize'. This is not my thing in the moment. What I create is the paths towards that. (Interview 4–6)

When the plan to set up a company was conceived, the mentor and place shaper did not perceive industrial corporations as innovative partners for investment and research cooperation. Instead, at the time, the late 1990s, VC funding seemed the more attractive choice.

The scientists who work in such companies have grown up with certain traditions and not necessarily with vision. The first thing they are asked is, 'How do you find this idea and in which direction is it evolving?' And I would guess that the answer was frequently, 'Nice to have, but we do not believe that this will be self-sustaining someday'. Scientists are not visionaries. They are down-to-earth. [*Venture capitalists*] could see something in this and at the time, they also had the means to really take a risk.

This was a risk business, but a risk business with a very innovative strategy, geared towards a specific market niche. (Interview 4–6)

While this assessment cannot be judged for its 'truth', it is noteworthy that it stands in contradiction to appraisals provided by actors in other cases such as case 3. The fact that the founding team has a strong propensity to work in an independent, self-determined manner may have added to the general preference for VC over an industry collaboration. In case 4, BIOMARKER, the young company would retain strong local ties to the Max Planck Institute. For example, its scientific base was developed by employing PhD students who were supervised at the institute. This early phase was thus characterized by a decidedly non-industrial style of working. As stated initially, this innovation received its first major investment from a Munich-based German VC fund, a fund set up by a leading German bank, in 1998. A representative of this VC fund who was involved in the investment at the time confirmed that in the late 1990s German venture capitalists were willing and able to invest in visionary ventures, even if scientific and commercial proofs were thin. The investors saw a potential in the patent situation, which promised the opportunity to establish a new technological paradigm and thus achieve a domineering market position – possibly on more than one market.

[*Company*] had a revolutionary idea: going one level above genetics, a 'loftier level'. The epigenetic level switches genes on and off. For example, cancer genes are switched on, which are supposed to be off. It has a use for diagnosis, perhaps also therapy; a great tool. You don't have to sequence the entire genome, only the switches. It allows for early diagnoses, a very brilliant, fascinating idea. I have to say it clearly, this elated [*people*]! (Interview 4–7)

The founding team was very young (a circumstance highlighted by later collaborators), and, except for the case's leading company founder, did not have an entrepreneurial track record. The mentoring department leader accompanied the team to face-to-face meetings with investors to vouch for the scientific validity of the pursued technological approach, helping it to be perceived as sound. The founding team was recognized as well balanced, since it included not only carriers of scientific expertise, but also technical and business expertise. With the mentor's help, leading international experts were enrolled as supporters. The patent situation was favourable. Beyond that due diligence, procedures were described as less strict than they are portrayed today.

What you can say for sure and generally is that, compared to 10 or 12 years ago, certainly today companies are only funded when they are more advanced in their development. As I said before, back then there were a lot of companies, talking about genomics or proteomics, which came directly from basic research, and which were financed on the grounds of a certain hype; because people had the expectation, once I understand genomics, the gene composition, I can address a lot of illnesses and solve problems. This turned out to be wrong. (Interview 4–7)

From a later perspective, the investment was seen as unusually daring – untypical for VC in general and German VC in particular. Therefore, the engagement is only understandable when the exceptionally enthusiastic atmosphere in the late 1990s is taken into consideration. In case 4, BIO-MARKER, the expectations of future returns on the side of investors were so optimistic, that building a company from scratch based on a radically new technology seemed realistic.

In the absence of established entrepreneurship infrastructures, the company established itself in an unused inner-city rear courtyard building. The building was transformed into lab space in handicraft by the entrepreneurs themselves. Equipment was partially created through DIY. PhD students from the 'mother' institute were employed as scientists in a joint framework with the institute, making the company appear less like a business and more like a scientific exploration platform with a wide array of potential applications (including cancer therapy). The company even attempted to engage in ground-breaking epigenetic research equal to the – still running – Human Genome Project (HGP).

> This scientific spirit led us to try to apply the technology in all kinds of fields. I always called it the Humboldt University at [*company name*]. Somehow, the difference wasn't all that visible at times. We did projects, you wouldn't believe it. Once we attempted, as a company, along with some public institutions, to more or less initiate the Human Epigenome Project; as a mirror image to the Human Genome Project, but not focusing on the DNA sequences, but on the methylation patterns. [...] I guess that back then [*name of leading founder*] saw himself as the second Craig Venter. 'Let's just do this!' We did start it. We did start it. (Interview 4–1)

Life and decision making in the company were described in images reminiscent of a 1970s commune or a shared student flat. Occasionally, VC representatives would join the nightlong discussions. The VC investors helped create this situation, but they would also initiate substantial changes, which would foster a stronger market orientation.

While the Berlin start-up with its vague business model and ambitious technological promise received funding, a different effort in the USA was less successful. A team of seasoned clinicians and experienced entrepreneurs in Seattle was, simultaneously to the work undertaken at the Max Planck Institute in Berlin, in the process of starting a methylation company with a clear focus on cancer diagnostics. This company would later play an important role in the innovation process. This case is the only one in which two localities *independently* accommodated an experimental realization of an innovation's key idea. One of the American founders explained:

> Well one of my good friends headed the Hutchinson Cancer Research Center, which is a national cancer centre here in Seattle that I think at the time was one of the largest cancer research centres in the United States. And he headed that for 13 or 14 years and he retired, and when he retired he had an idea to try to develop diagnostic tests to detect cancer at very early stages. And over time he and I talked about this idea and we got in contact with another one of his fellow cancer centres' directors and travelled down to Southern California [*to meet*] one of the world leaders in epigenetics. And this fellow, [...] he's the one who really guided us toward this epigenomic or specifically DNA methylation as a potential for developing tests to detect cancer at very early stages. (Interview 4–4)

As the quote shows, this venture was much more centred on a particular application and business model. However, despite a proven business case, excellent scientific credentials and entrepreneurial track record, no VC fund was ready to invest in the idea. The team found the following explanation for their lack of success: The business model was too conservative – diagnostics rather than therapeutics, specific products rather than a big technological turnover – for the heat of the late 1990s, a time of hype and exaggerated expectations. Later, an organizational integration of both ventures would ensue.

3.2 University hospitals and other early innovation ecologies

The remainder of the cases studied add further facets to the overall picture but confirm the main dynamics of agency. They also serve to add an emphasis on the role of public investment, including public venture capital, in early stages of biotech innovation. In case 2, SYNTHESIS, a peptide chemist worked in a government-funded yet industry-oriented research facility[2] in Braunschweig in the early 1990s. In this organizational setting, in contrast to Max Planck, patents and applicable solutions were highly regarded, which alleviated the need for mentorship for application-minded researchers. The potential conflict between different epistemic communities, whose meeting is typically a precondition for innovation, was likewise moderated by organizational structure. Chemists routinely synthesized molecules which were used for research by biologists on site. In this constellation, the differences between both epistemic practices were not experienced as conflict-ridden, but as stimulating.

> It is a field where chemistry and biology come together so to say. [...] We were chemists, we made peptides. That is, we made tools, which were important for the biologists who did immunologic research here at the

institute. And in exchange with the more biology-oriented research projects we developed methods [...] which could advance this research. We picked up problems the biologists were working on and developed new methods to advance these efforts. And that was always fun when we worked in an interdisciplinary way. (Interview 2–2)

This narration of 'peaceful' collaboration stands in sharp contrast to other accounts, which highlight the difference in epistemic practices, particularly their time structures, between chemistry and biology. These differences are deeply rooted in the materiality of, quite literally, their subject matter:

Chemists have zero understanding of biology, zero! For example, if we breed cells to produce a certain protein. The cells have a certain rate of division, and it simply takes a certain amount of time till you have enough cells, which produce enough material. That depends on the division cycle of the cells. 'How long will it take for you to produce this product?' I answer, 'That'll take me six months'. Then they ask me, 'Why, do you only have two people on the job? Why don't you deploy six?' 'That won't make the cells grow faster!' I continuously had to put up with this kind of argument. (Interview 2–3)

The frequent shifting between on-site engagement with 'customers' in biology, discussions about different methodological approaches (experimental, deductive/combinatorial) in his own field of peptide chemistry and quiet contemplation allowed the scientist to come up with a new idea.

In most cases you have the idea when you are in a different place entirely, possibly in bed or wherever. [...] We have established many variants and methods which ran into dead ends at some point, because they simply were not technically feasible or practical. It is a very arduous process, and often when something does not work straightaway, you make it more and more complicated, because you want it to work. But in this case, it was back to a very simple idea, which then worked from the start. (Interview 2–2)

The scientist invented a method to generate large numbers of molecular combinations and test their interactions with other substances experimentally. The inventor drew on experiences from his earlier work in peptide chemistry, including his PhD project, when he made the invention. In its most essential form, the procedure involved a sheet of paper with a matrix on it and a person who would pipette substances into the individual fields after a predefined protocol. This activity was both mind-numbing and prone to error. To remedy this limitation, the chemist had an engineer construct a synthesis robot. This engineer had, earlier in his career, worked with peptide chemists and his company had sold synthesis robots to the chemist's institute before.

The emergent automated assembly could easily be moved from site to site. It became a mobile boundary object between peptide chemistry and immunobiology and as such helped bring representatives of both communities together.

In this case, the inventor and the entrepreneur who would commercialize the invention were two different individuals: When another peptide researcher, who had just set up a working group at Charité university hospital in Berlin, learned about the technology during a presentation, he could easily spot the potential for economic exploitation. By describing himself as an entrepreneur with a taste for the challenge of company founding, he set himself apart from the more science-oriented original inventor.

> This was the point, if you will, that the inventor, whom I know well and with whom I collaborated a lot, was absolutely not an entrepreneurial type. And in this respect, I had a different orientation than he did. (Interview 2–1)

He bought several synthesis devices and used them to set up a service-oriented peptide synthesis company in Berlin. To this end, he also acquired the patents protecting the invention, giving licenses to the original inventor and the engineering company. Returning from a career station in HIV research in the US just a few years earlier, the founder had decided to set up shop in Berlin because the city was in an exciting and dramatic process of transformation (a very personal locational preference). Academic entrepreneurialism had not yet been established as a socially accepted career choice (the BioRegio competition had not yet taken place), so the decision to set up a company was a clear breach of an existing order. The entrepreneur relished this challenge. He also rejected the later proliferation of founder support because to him it banalized the idea of entrepreneurialism and produced unviable businesses.

> You cannot imagine that today. Back then, it was completely stigmatized to found a company. The professors at the time frowned on people who founded companies. You were not supposed to make money. All that changed suddenly one or two years later when this BioRegio competition came. Then you had to found a company. It was chic. And there was a situation that people founded their companies at their universities [*author's note: i.e. they remained in the university context*] and it all became a mess. This is why we were successful. For the people who founded later, it was all too easy. Money was thrown at you. Every professor founded a company. For one reason or other it did not work out in most cases. (Interview 2–1)

When the company was founded in Berlin, there were no formalized support structures. In this situation, the founder received support, encouragement and mentorship from a Berlin pharmaceutical company's head of R&D, who would later become the city's cluster manager for the health economy. The company was to deliver synthesis services for tailored organic molecules. With this comparatively conservative business model, the company did not require venture capital. Only in 2001, when a shift in the business model occurred, the company's VC history began.

University hospitals as places to start from or places to go to appear in other cases, too. The remaining three cases illustrate medical biotechnology innovations which emerged in university hospitals in different times (early and late 2000s). In case 6, STEM CELL, a team of medical researchers at Cologne university hospital, led by a geneticist, founded a company in 2002/2003. The initial founding idea was to develop gene therapies. The founders obtained small-scale seed financing as well as consultation from Technologiebeteiligungsgesellschaft (Technology Equity Company, TBG), a public, national technology financing body established in the 1990s. The choice to approach TBG was made because one member of the founding team had previously participated in a company founding, in which TBG had also been an investor. TBG was based in Bonn, near the founder's location. TBG financing at this time was described as very readily available. As one executive explained:

> And what came into it very early in 2002 was a loan from KfW. They were these typical bonus shares, which KfW shed out via TBG so generously [...] over all possible and impossible ideas. A great thing. On the other hand, also an unbelievable structural error. Because you create an incredible amount of ideas, which all collapse at some point, because successive financing is missing. Well, they see it in a Darwinist way and say, 'The fittest survive. And for the others it's just bad luck'. (Interview 6–1)

The fact that TBG did not act like a classic financial investor is illustrated by the circumstance that gene therapy at this time was deeply discredited in the VC scene and in the public. Soon the founders realized that the initial plan was unfeasible for technological, market-related and IP-related reasons. So, following a process of heated discussion, both the technology and the founding idea were swapped. While failure and reorientation occur in other cases, too, this is the only incidence in the cases studied, in which an investor allowed such a dramatic reorientation. Another invention already made, which was more reliable and better protected by IP, was put forward: The ability to make human stem cells immortal, that is, able to divide indefinitely. This technology was now to be used to create human proteins for therapeutic purposes.

> The human cells produce exactly the proteins, antibodies and other biologic drugs that the human body would produce itself. And not like a rat or a duck or whatever would produce. There are differences. The

differences are such that, if you were to inject proteins from a duck, they would not work as well and there would be side effects. In particular, there are, as they are called, immunologic incidents, i.e. rejection; side effects which are simply allergic in nature. (Interview 6–1)

Thus, in this case early investment preceded the identification of an exploitable technology. This occurrence was described as highly unusual by respondents, yet not fully untypical for the institutional context of its time. While the private VC market for biotechnology in Germany had essentially collapsed during the burst of the dotcom bubble in 2000, public financing institutions were still running in the expansionist mode of the late 1990s. However, this changed soon after. Case 7 CANCER IMMUNE provides an interesting counter example in a similar field (gene therapy, genetic vectors) several years later. Yet, it shows that early investment can still be an element in an idea formulation process. In this case, a researcher experienced scientific developments surrounding gene therapy as a member of both a Max Planck and a Helmholtz working group in Berlin in the early 2000s. In a project funded by the BMBF, which funds application-oriented research, he studied how viruses could be used as carriers ('vectors') to bring genetic material into cells and thus change their DNA. He also sought a realistic field of application for the method and found it in cancer therapy: cancer cells could be infused with genes which counter their inherent ability to put surrounding immune cells 'to sleep'. Under the influence of the new genes, the cells would actively stimulate and attract the immune system and thus commit, as it were, assisted suicide. This decision was the result of discussion processes within the scientist's research team.

By relocating to Hamburg Eppendorf university hospital, he found an environment with all the relevant sites and practices to test this idea. The environment became particularly relevant as it allowed the future founder to 'regroup' after a disappointing encounter with oncologists (the proposed users of the idea) and evade them via a more supportive community. The oncologists had safety concerns and did not appreciate viruses as therapeutic tools.

We talked to many clinicians about the field of application. And we went to the oncologists, of course. With cancer, you go the clinical colleagues in oncology and say, 'So, what would a drug have to look like?' And they say, 'Gee, a virus [...], a modified cold virus as a vector, which produces these immune hormones, and then what happens?' An oncologist cannot imagine how this [...] affects the whole body, how it activates the immune system. [...] We then got together with vaccine developers, and with vaccine developers you get viruses without end. They create virus-based vaccines, partly live, partly weakened, no problem at all. [They said,] 'We can see this, great, it will work'. [...] The moral of the whole story is: in innovation, in clinical practice, in research, an incredible number of languages are spoken, and you have to find the people who speak the same language as you. And if you enter via the immune therapists you can get traction on the same invention, the same technology, while in the wrong, or primarily non-responsive field, people will not understand you. (Interview 7–1)

Later, he was approached by a private investor who encouraged him to commercialize the idea. With a small team, he set up a company on the university hospital's campus. Surgeons and oncologists later readily participated in the advancement of the idea. As in other cases, evading an inhibiting environment was equally important as finding a responsive one.

In this case, place shaping did not occur as the agency of a leading figure in science. Rather, a private investor or 'business angel' acted as first investor, mentor and place shaper. This individual's place shaping activity was not limited to a particular locality but focused on specific sites (medical research sites at universities) within a territory (Germany). Having gathered experience as a university professor, a company founder and a consultant to the pharmaceutical industry, she had decided to counsel German universities on how to make their newly established BSc degree programs more relevant for professional practice. During her frequent visits to universities, she encountered numerous scientists who pursued innovative ideas and had a general interest in company founding but did not have any experience with the process of founding. She began to mentor and coach such scientists *ahead of company founding*. She used revenues from her own earlier entrepreneurial activity to invest in the emerging start-up companies on a small scale. The idealistic motive of helping creative scientists put their ideas into practice was her key motivation, while the hope that her investments would produce return was a secondary one.

Around the year 2005 she approached the scientist at Hamburg Eppendorf university hospital, encouraged and supported him in the creation of a start-up company, and made the first investment. While the entrepreneur used other support structures (such as local business support, national applied research funding, business plan competitions and networking events), the investor supported him in making strategic decisions. Building on the strategic leeway created through the first investment, the entrepreneur managed to win High-Tech Gründerfonds (High-Tech Entrepreneurs Fund, HTGF) as a second investor. While the rationale behind this public VC fund is that by creating relational proximity to private VC funds, such can presumably be better attracted as co-investors, in this case HTGF would remain largely passive on strategic matters. Instead, the further development of the company was crucially influenced by the business angel's support and mentorship.

Case 8 NEURON, finally, is exceptional in the sense that the key entrepreneur driving the idea forward does not rely on any form of mentoring or 'place shaping' work by another individual. He does, however, rely on established networks and boundary-spanning experience. The case also provides yet another example of the need for evasion and reorientation in early innovation stages. A professor of medicine with a double affiliation in Germany and the United States was in the process of evaluating the performance of an Alzheimer's drug, which had been in the market for several years, for a large US pharmaceutical company. He found that the drug had extreme side effects which often rendered it impractical in therapy. The active ingredient had been used for generations by people in Eastern Europe to counter the cognitive effects of ageing. When it came to the market as a natural product, heavy side effects set in Western European and North American users.

> We had not expected this, I have to admit. And likewise, the people who had used this substance for centuries as a folk medicine against maladies of aging, in Bulgaria, South Asia, South Russia, they don't have those side effects because they got used to it, as it were; in the same way that there are arsenic eaters among mountaineers, who can take really a lot of arsenic, dosages which would kill other people, but who can do this for better oxygen absorption and who have gotten used to principally lethal dosages over the years. (Interview 8–1)

His relationship with the pharmaceutical industry was based on a long history of exchanges centred on his field of expertise: a certain type of receptor, found in the central nervous system, whose presence and frequent stimulation appeared to be relevant for nerve cell survival during neurodegenerative diseases. Nicotine and related substances stimulate this receptor. The drug in question belonged to this substance class. The entrepreneur had both a research group at Mainz university hospital, his home institution, and a network of peers at his disposal. Furthermore, he had participated in the development of his region's BioRegio strategy and participated in international standardization boards at the intersection of academia, industry and governments. Thus, he had the seniority and inter-institutional reach associated with 'place shaping' agency.

One long-standing cooperation partner, a Vienna-based scientist, possessed a patent for a chemical synthesis process for said substance. Embedded in this constellation of exchanges, he came up with the idea to perform a chemical alteration[3] on the substance that would allow it to cross the blood-brain threshold more easily, and thus drastically reduce the side effects while increasing performance. However, the pharmaceutical company had no interest in bringing this new variant to the market. According to the entrepreneur interviewed, this was based on the consideration that the development of an improved version would damage the market position of the existing product, which was still patent protected for a considerable amount of time. The inventor perceived this as an opportunity to set up his own company and pursue the idea.

The company founder enrolled a consortium of public semi-public investors as initial investors in 2005. HTGF acted as first investor and was followed by a regional public–private VC fund as well as KfW (Reconstruction Loan

Corporation), who provided co-funding. In this case the apparent simplicity and marketability of the proposed project (improving existing pharmaceutical products) was described as decisive argument for the investment by one of the investors. However, the company failed to obtain further VC investment after the original public/semi-public investment. According to the founder, the technological approach upon which the company was founded deviated from the dominant school of thought in the field of pharmaceutical product development for neurodegenerative diseases. He provided the following narrative: In the absence of a reliable 'initial public offering (IPO) window', venture capitalists focus on industrial corporations as the only promising exit option. In doing so, they concentrate on technological approaches which are supported and pursued in the industry. As a consequence, proponents of scientific minority positions do not get VC funding. Although this assessment is difficult to verify, multiple respondents stressed the negotiation power of industrial corporations vis-à-vis venture capitalists in the absence of a stock-market-based exit option, and consequently the dependence of VC investors on industrial strategy. This assessment of the financing environment in the mid-2000s is markedly different from what respondents described as the financing environment of the late 1990s. In case 8 NEURON, the innovation stalled for several years before it picked up again around 2010.

3.3 Founders, mentors and idealistic investors: key dynamics in early stages of biotech innovation

In conclusion, the early stages of biotechnology innovation are characterized by typical dynamics, which nevertheless play out in specific ways in each case, depending not least on time and place. The term 'early stage' here is understood as encompassing the development towards the formulation of an idea, its explication as well as the first steps on a path towards commercialization, which themselves are experimental in nature. The result can be characterized as a highly localized, open, instable and hybrid actor-network which manifests a new technology for the first time outside the purely academic laboratory. The 'technology' at this point is a repeatable yet not industrialized techno-scientific procedure supported by inter-connected artefacts, competent people pursuing epistemic practices, located in a supportive physical place. It is embedded in the organizational form a company which, however, does not work like a corporation, but rather like an inter-institutional hybrid driven by individual enthusiasm. It is furthermore embedded in a highly place-specific environment, typically close to an academic mother institution and within reach of routinized daily mobility. External supporters like academic advisors and investors stabilize and legitimize the assemblage in its institutional environments.

Some key ideas from the theory of translations can be applied to characterize the emergence of this initial technology assemblage, such as the notion of problematization: The actor-network is organized and ordered around a key idea, which also represents a set of meaningful relationships. Their meaning lies in their capacity to overcome an 'obstacle problem' shared by all participants, typically a deficiency in a techno-scientific practice, which leads to unsatisfactory results. This problem is claimed to affect both users and the scientific community. While users are portrayed as suffering an immediate, often physical disadvantage, the scientific community is affected in terms of its identity, credibility and societal relevance. Interestingly, in the cases studied the proposed solutions were formulated solely in scientific terms and did not draw on user practices as potential resources. Also, the original entrepreneurs were represented *exclusively and personally* as the providers of a specific solution, as a short recap of case 1 shows. The problem formulation in this case reads: Existing Encapsulations of biopharmaceutical active ingredients are inelegant and cause unnecessarily invasive treatment routines. A new Encapsulation method is the solution. Only we, the founding team, can do this (case 1, ENCAPSULATION).

> The product itself is from [*company name*]. [...] It's a peptide. But they made a formulation which is really terribly bad. Yes, it's like a medieval torture method. [*It's*] For horses, not for humans. And we've created a really beautiful formulation now, which we are going to license. (Interview 1–1)

As in Callon's (2007) story of the scallops in the Bay of Saint-Brieuc, the relational work of 'interesting' and 'enrolling' actors under a shared mission is a hybrid activity – directed at both human and non-human actors. It is an ordering activity which produces both new relations between 'social' and 'technological' entities. For example, in case 3, GENE FUNCTION, the combination of genetic methods and a biochemical observation technology for cellular metabolisms reflects – or rather produces – an overlap of these two epistemic communities (genetics and biochemistry). A new biotechnological procedure is therefore 'naturally' embedded in a scientific environment. A further element which featured strongly throughout the cases was the importance of an enabling technology based on IT and engineering: By creating a network of procedures and artefacts which 'runs' on IT and are pieced of machinery which were either newly created or imported from other contexts, a hybrid linkage was established not only between different epistemic communities, but also between scientists and engineers. Users on the other hand are either absent or only indirectly represented in the situations, in which the problematizations are formulated. In some cases, user practices served as sources of inspiration, for example, those of clinical experts in university hospitals or industrial product developers. In other cases, users were represented by epistemic communities which have strong linkages to application, such as immunology or oncology. Hence, there frequently was a presence of what might be called surrogate users or proxy users. In Ibert and Müller (2015) this form of interaction is described as the 'hanging out'-relationship. But in most cases, the innovation processes developed away from these user encounters. Neither users nor user practices featured strongly in the problematizations proposed by entrepreneurs, except as recipients.

Throughout the cases, the choice of business model was an incremental expansion on what was there already. It was strongly informed by entrepreneurs' personal identification with the technology and their take on its potential practical value (not its market value) as well as by scientific rationalities and value appreciations. Typically, entrepreneurs had a strong inclination to include the development of an end-user product in their 'bouquet', to *personally* realize a practical impact. These individual developments in most cases were embedded in some form of technological context: a small-scale, experimental technology platform or a pipeline of various applications (see next chapter). This embedding reflected the entrepreneurs' appreciation for the organic interdependence, as it were, of all the things which could be done with a particular procedure. Crucially, commercialization model decisions were not made following a clear-cut market or business logic. Entrepreneurs strived to make a business-case for their respective ideas as a matter of practical necessity (e.g. enrolling early investors), but the technology-oriented definition of the project came first. Founding teams, however, varied somewhat with regard to the strictness, with which they adhered to a particular model. In several cases, entrepreneurs were ready to be led on a particular commercialization path by industrial partners.

Throughout the cases, a creative tension was visible, as the cited examples show: Creating the early stage of an innovation process rests on key dynamics of agency; not one, but several interacting agencies. I identified two core types of agency, as introduced at the beginning of this chapter: *Enthusiasm-driven entrepreneurship* and *place-shaping*. Key characteristics of the two – which in most cases can be linked to concrete individuals – will now be discussed, before advancing to the role of investment later. Crucially, the two forms of agency connect orders of worth with situations of knowledge work, of knowing and not knowing in practice, and assign value to epistemic as well as entrepreneurial endeavours. They mark out opportunities in local and temporal conditions and set out to exploit them.

'Enthusiasm-led entrepreneurship' refers to the development and vigorous pursuit of one focal technological idea – often by a leading entrepreneur henceforth denominated 'enthusiastic founder'. In the process, collaborators and supporters are enrolled, all based on highly individual, intrinsic motivations, and centred on one person or a small team, who drive their project forward. In this case it is comparatively easy to connect an action logic with a concrete individual. In each case studied, a biotechnology company was founded. Most founders roughly match the abstract image developed here. It resembles the disruptive, Schumpeterian entrepreneur (Schumpeter 1997 [1911]). The founder combines a strong sense for opportunities to create new things with the ability to inspire and convince others, typically in face-to-face interaction, hence, charismatic leadership (Weber 1980). Enthusiasm was recognized as a driving force in user-induced innovation (Brinks and Ibert

2015). In the strongly science-based field of biotechnology, analytical aptitude may be considered the more relevant source of novelty. However, enthusiasm, i.e. a very personal identification with an idea and a strong emotional impulse to implement it, is also an indispensable resource of innovation. The ideas in question were found in environments of scientific practice.

In terms of orders of worth, enthusiasm-led entrepreneurship is guided by a) a strong and highly individual identification with the practices cultivated in epistemic communities as well as b) a high valuation of recognition beyond the boundaries of one's peer community and also for realizing a practical impact or 'making a difference'. The earlier translates into a willingness to participate in the revision of epistemic practices and the redrawing of community boundaries, however without the desire to delve deeply into the politics of professional identity and the associated heavy conflicts. Enthusiastic founders experience the tensions at the overlap of different epistemic practices (and also practices outside the realm of science) as something creative and seek out opportunities to create new approaches out of this tension. The latter translates to a propensity to search for opportunities to break free and do one's own thing. In all but one of the cases studied, these two aspects of the logic were 'lived' by one individual. These individuals were scientists. While most were further advanced in their careers (postdocs or professors), one was a PhD student. They worked in public research facilities, again with one exception, who worked in a pharmaceutical company. These entrepreneurs typically became company founders or participated in the founding of companies.

They related strongly and personally to the ideas they proposed and the projects they initiated. The personal identification with an idea typically outweighed the identification with the orders operating in communities, institutes or companies (except their own start-ups). The ideas were typically legitimized and underpinned with references to the public good, to a generally unsatisfying state of affairs which needed to be remedied. Thus, enthusiasm-led entrepreneurship comprises a high degree of idealism – in combination with a strong desire to be recognized. Founders perceive civic responsibility as an important value. Furthermore, they perceive this type of value as complementary to the value of building individual reputation beyond the scientific peer community by creating an applicable solution to a problem. The bureaucratic orders and highly structured career paths of the academic system as well as large corporations are viewed with more critical distance. Crucially, the market order is seen as instrumental to the other two orders of worth (civic responsibility and individualized fame), but (in most cases studied) not as normatively valuable in itself. Concepts and notions of market strategy and market success are less frequently cited by enthusiastic founders. Instead, the market order rather appears to be a necessary evil for the higher goal of establishing an individually attributable idea in an environment of application (see Boltanski and Chiapello 2007).

Throughout the cases, founders typically acted in situations in which solutions were sought for scientific problems which had an established or at least assumed linkage to application. These search processes provided opportunities to rethink epistemic practices and the boundaries between epistemic communities. In two cases, new epistemic communities were being created at the time and in conjunction with the concrete action situations in which the founders found themselves. The reason for this disruptive dynamic is that each provider-user linkage (e.g. an academic scientist working as a service contractor for a pharmaceutical company) also implied a fixed relationship between epistemic practices. For example, a certain branch of chemistry would typically provide molecular synthetization services to biologists. The arrangements of 'service provision' between epistemic communities played out in inter-organizational or intra-organizational constellations. They were sometimes institutionalized in the form of research programs and funding schemes. In innovation processes, these relations with fixed role definitions for human and non-human actors (i.e. existing translations) ran into trouble due to different forms of dissidence, or because competing translations were pursued by other actors. Enthusiastic founders recognized and actively pursued these changes.

Throughout the cases, entrepreneurs identified strongly with the epistemic practices in which they were enculturated. Their epistemic cultures provided them with a sense of value for specific forms of knowing in practice and a propensity to 'do it right'. This refers to habits of thought as well as routines and intuition in human–non-human interactions in the lab. The normative aspect of epistemic culture, which deems some kinds of epistemic enquiry worthy and relevant, while deeming others not worth pursuing ('negative knowledge', Gross 2010; Knorr Cetina 1999) also became apparent. However, they employed these aspects of their identity not to defend a coherent state of the art, but as a resource to criticize the state – either within an epistemic community or with regard to the way epistemic practices related to each other.

Hence, enthusiasm-led entrepreneurs are involved in acts of dissidence against the 'orthodoxy' represented and defended by often high-ranking members of epistemic communities. How is this done? Interdisciplinary training frequently occurred as a factor in enthusiastic founders' professional identity. Many occupied temporary positions of 'multiple insiders' (Vedres and Stark 2010) or at least 'immersed' (Wenger 1998) in the practices of neighbouring communities. Typically, ideas were generated because of such boundary-spanning encounters. By capitalizing such ideas, enthusiastic founders acted as 'tertius gaudens'-type brokers (Klagge and Peter 2009). In doing so, they also participated in a redrawing of epistemic boundaries; a process which has can be described in terms of its manifestation in time and space. The most general statement is that this process is characterized by a high degree of mobility, but also by a need for continuity in the form of stable, localized settings. Enthusiasm-led entrepreneurs engage both with practices and practitioners from other communities and from their own. While discursively engaging with practitioners can be done in settings of temporary co-presence (i.e. workshops and conferences), engaging in a practice requires that one is present in the site where the practice is carried out.

To experience the fine points, the accidental incidents and serendipitous encounters, trials and errors, disturbances and imponderabilia of the practice, but also to understand the way practitioners evaluate and appraise results, this presence requires some regularity. It needs to be embedded in one's day-to-day action space – possibly, but not necessarily, in the same site (e.g. an interdisciplinary research institute), often in the same locality. Such encounters can be experienced as positively reinforcing if the other practitioners are supportive and willing to be enrolled as allies, but also as challenging if they refuse or actively oppose the entrepreneurs' proposals. Both support and refusal can be found within one's community and in other communities. Crucially, both experiences can be considered inspiring. Enthusiastic founders connect both experiences in their mobility. In addition to engagement, they also make use of retreat, the possibility to quietly reflect, read, experiment individually or exchange thoughts with very close allies, often friends and family. This aspect requires a degree of sovereignty over time and space in one's everyday activity pattern. Over time, this combination of friendly engagement, conflictual engagement and retreat enables enthusiastic founders to evaluate observations differently and perceive opportunities to approach concrete problems in a radically new way. In terms of quantified research output, their contribution is not necessarily exceptionally big. The real contribution is twofold: a) creating a shift in perception through relational work, seeing an opportunity for a new combination; and b) going through with the idea on a path towards application, a path which is highly uncertain.

Arranging these experiences and actions in a professional biography is in part done by sequential mobility: moving from one station to the next within a career and occasionally taking a time-out. It is also achieved by frequently visiting different sites of practice in the same time-span. But enthusiastic founders at some point also seem to require a stable setting to develop ideas. This setting on the one hand involves a location where different sites of practice (private and professional) are located near to one another and can be easily and quickly connected in everyday life. This location affords diverse opportunities for engagement and retreat. The setting, on the other hand, also involves an organizational environment with an order which legitimizes and encourages the re-negotiation of community boundaries. Additionally, resources are required to change and adapt sites to a new practice. If enthusiastic founders have such a setting at their disposal, for example by occupying a chair at a university hospital, they exploit it.

If such a setting is not available to them, they search for it. Since what is sought is an experience of difference, the search is by definition 'non-local'. It is also open and opportunity-driven; enthusiastic founders keep their eyes open for and are receptive to promising, often coincidental encounters. In addition to looking out for desirable circumstances, the search can also follow the explicit desire to escape an inhibiting environment. In terms of material space, the 'base' sought is a location (a city or a region), which provides a specific arrangement of different sites. However, search and mobility are also guided by very personal motivations: looking for a vibrant place to live and work or moving into the vicinity of family members. Enthusiastic founders carry their ideas with them, but this embodied mobility of knowledge is complemented by other mobile elements. On the one hand, the pioneers often operate in an environment where new devices and ideas have reached a certain proliferation. Consequently, the sites and locations they seek out provide interfaces and elements which can be creatively recombined. They are receptive. On the other hand, enthusiastic founders also participate in the creation of new, mobile devices and procedures, which can be viewed, engaged with and tested in situ.

As shown in the previous section, enthusiastic founders depend on environments in which they can develop and pursue their ideas. The environments which provide such opportunities are not simply excellent research institutes or business incubators. They are relational spaces, situated in specific localities and in specific, limited historic time periods. Within these, ordered relations of human and non-human actors are opened up, so that an entrepreneur has the freedom to reinterpret them. Such spaces emerge when and where innovation-friendly 'place shaping' happens. The term 'place shaping' is used in the context of local government and urban design as relational, process-oriented and governance-orientated modus of leading and effecting local development (Carmona 2014; Madden 2010). I use the term in a more specific way to denote the gradual shaping, particularly opening-up of localized innovation ecologies. The key actors, who were identifiable in all cases except one, will be described as 'place shapers'. They are typically high-ranking individuals in one institutional environment and thus have substantial resources – money, reputation, relationships – at their disposal (see Table 3.1 for an overview of place shapers in the studied cases). They use these resources to renegotiate boundaries and create opportunity spaces. The department and institute leaders referenced as key enablers in almost every case provide the basis for the following theorizations.

Conceptually, the place shaper can be situated in proximity to the institutional entrepreneur (Crouch 2005). Yet, place shapers often have a history of 'classic' entrepreneurialism, i.e. as company founders, which preludes their activities directed at changing the institutional environment. In doing so, their reach is smaller and their actions less potent than those of institutional entrepreneurs. They do not change the institutional set-up of an entire field (Fligstein and McAdam 2012). Instead, they create localized and temporally limited 'bubbles' in which orders are up for negotiation, and which of course can contribute to overall structural changes in the long term. This is especially the case when they manage to establish a new organizational or institutional form to 'back up' their own local interpretations and practices. Consider for instance the institute leader in case 3 GENE FUNCTION, who established a site of practice, which clearly had a basic research mission, but accommodated elements of industrial working culture. Place shapers are clearly 'multiple insiders' (Vedres and Stark 2010), often with long-standing

Table 3.1 'Enthusiastic founders' and 'place shapers' across the cases

Case	'Enthusiastic founder'	'Place shaper'
1 Encapsulation	Scientist in a pharmaceutical company	Cluster manager, former founder and Gene Centre leader
2 Synthesis	University hospital working group leader (also partially the case's 'Pipeline builder', see Chapter 4; in this case the inventor is a separate person)	Pharmaceutical executive, among others
3 Gene Function	Post doc at Max Planck Institute, among others	Max Planck Institute leader
4 Biomarker	PhD student at Max Planck Institute	Max Planck department leader
5 Autoimmune	Researcher at Max Planck Institute, among others	Max Planck department leader; additionally a pharmaceutical company executive and former founder
6 Stem Cell	University hospital scientist, among others	University hospital working group leader
7 Cancer Immune	University hospital scientist	Private investor, founder and consultant, former scientist (also the case's 'Pipeline builder', see Chapter 4)
8 Neuron	University hospital scientist working in a collaborative relationship with a pharmaceutical company (also the case's 'Pipeline builder', see Chapter 4)	

Source: Own design

affiliations in academia and industry, academia and policy making or business support, or similar combinations. Beyond those immediate roles, place shapers act as mentors (protecting, encouraging and counselling nascent entrepreneurs within their domain), but also as brokers (building linkages to other domains such as investment). The particular form of brokerage – building lasting links and benefiting from the emergent ecology – has been denominated 'tertius iungens' brokerage (Klagge and Peter 2009).

The locality, that is, the presence of sites and actors within the reach of everyday mobility, has an important influence on the process of place shaping since the latter involves coincidental encounters, 'under the radar' activities, informal low-threshold negotiations, personal relationships and preferences as well as life cycle dynamics beyond the professional biography (consider the informal re-negotiation of IP-related rules in case 5 AUTOIMMUNE). Place shapers are pragmatic. They value novelty in their own domains. Like enthusiastic founders, place makers like to see it when new knowledge is applied to an end which serves society. Therefore, they appreciate academic entrepreneurialism and industry collaboration. However, they have no interest in pursuing particular ideas or projects for themselves. Instead, they value such initiatives among their colleagues and team members. Supporting them is a way to exert influence indirectly and to 'softly' enrol more actors into one's own idea of knowledge production. While the set of orders of worth mobilized by place shapers resembles that enacted by enthusiastic founders, the instrumental relations between them is different. The relational work of mentorship (Ibert, Müller and Stein 2014) and 'tertius iungens' style brokerage (Klagge and Peter 2009) is a way to act upon this set of values.

The build-up of an experimental technological assemblage outside a public research lab requires funding which goes beyond research grants. After the funding of a company, entrepreneurs have the opportunity to apply for funding from public actors (such as BMBF) or by foundations, and in several cases, this took place. However, in all the cases studied, company founders also strived to enrol investors to fund the advancement of their technologies. The way in which investors were approached, interested and enrolled, and the way in which they in turn related to the entrepreneurs and their projects showed some repeating characteristics: Typically, the investors were supportive to the problematizations proposed by the entrepreneurs. Hence, they became allies. The early activities of investment in biotechnology innovation processes can, I would argue, be understood as *extensions* of the mentoring and place shaping activity. They rested on situation-specific relational proximities, pre-existent relationships and sometimes encounters based on physically proximate location. They were also strongly influenced by the cognitive framings and preferences of founders and place shapers alike.

Early stage investment is not a singular event involving one singular investor, at least not typically. Instead, it is a process of building an initial network of investors. In this process the very earliest investors contribute to identifying and enrolling further investors. The partial investments are still very small for biotechnology conditions (that is, adding up to total investment sums well below €500,000). Early investors are supportive to experimental technological assemblages in the sense that they tolerate that the company in which they invest is not 'investor-ready' in textbook terms. Instead, early investors participate in the building of localized hybrid assemblages of objects, people, practices and sites, which are still in the process of being defined and delineated. Typically, no realistic business model is present at this stage. Early investors tolerate the openness of this type of situation and also contribute to its eventual clearing up. Crucially, they can do this either by being active and knowledgeable or by trusting and being rather passive. In every case, they allow considerable leeway.

This process is highly place-specific. This means on the one hand that the pattern of presence and absence of actors in a locality has a very strong influence on the way in which investors can be met, identified, approached, interested and enrolled. In most cases, early investors were present in the respective city or region, either by co-location or by frequent co-presence. On the other hand, it means that the spatial practices and orientations, and the presence of both entrepreneurs and investors in particular localities and sites of practice (the latter is intimately connected to being knowledgeable and having experience in a particular technology field) usually overlaps prior to the investment. This aspect corresponds with previous interaction experiences and pre-existing relations.

The findings reflect Wray's (2012) findings regarding the spatial practices of VC investors and the relationality of investment opportunity creation: investors participate in the creation of 'investor-ready' companies, and they do so in a spatially differentiated way.

Early investors are thus a-typical in their conceptualization and appreciation of value and can be characterized as such in two ways: Firstly, taking the classical, capital-market-based, exclusively profit-oriented venture capital firm as a reference point, their expectation of profit and 'exploitation-readiness' is different. They can also be seen as a-typical if they act differently in one situation than they would at other times, due to very specific situational circumstances. Both types could be observed in the case studies. Most investors who could be enrolled had an intrinsic appreciation for the technologies developed, beyond their potential for financial commodification. In their action logics, they drew on other social orders beside the capital market while still retaining the goal to earn money. One such alternative logic is that of individual, partly altruistic appreciation for entrepreneurial activity based on their own experiences of entrepreneurialism and boundary spanning. This logic is typically associated with the figure of the 'business angel' (Harrison and Mason 2000b).

Another example for an a-typical (or not immediately capital-market-driven) logic of value appreciation is that of industrial product developers who strive to appropriate pioneering technologies and use them to improve their own market position. This logic is influenced by the goal to make profit and also by the need to refinance via the capital market, but it entails a different perspective on a technology. Instead of asking whether an investment object will be understood and will trigger interest by an unknown capital market player in the future, an investor asks whether the object is potentially useful to himself, immediately.

A third logic is that of technology investment as a means to promote the public good: to incubate businesses, to create highly skilled jobs and to carry new scientific knowledge towards application. This logic is typically associated with public VC funds.

In addition to structurally a-typical ways of assessing possible investments, investors were also subject to a very time sensitive, situational adaptation to circumstances. In one case (case 4 BIOMARKER), classic VC investors could be enrolled for a radically new technology during a time of unprecedented boom (the late 1990s), an event the actors involved agree would have been impossible to repeat at any point in time after the burst of the dotcom bubble in 2000. In the cases involving industrial investors, time was also a critical factor, as each investment depended on the opening of an opportunity window in the respective company's product pipeline.

The supportive role of early investors, however, is just one side of the coin. Early investors also are extremely important to the subsequent unfolding of innovation processes as they induce the changes necessary to enrol more 'normal' investors in the future and to manage the transitions necessary for entering a market. This role warrants close analytical attention. Early investors exert a very selective and discriminatory influence on the enrolment of subsequent ones. Their perceptions of interests and conflicts, of compatibility or incompatibility of investment logics, affect innovation paths. Regardless of this influence, early investors are still at risk of being crowded out, bought out or simply diluted by subsequent ones. This dynamic is important because many a-typical investors have a very sensitive and unstable relationship to biotechnology. The ways in which early investors experience interaction with more potent, later-stage investors is likely to influence their willingness to engage with biotechnology in the future. Hence, the drawing and redrawing of the boundary of biotechnology itself is affected.

Notes

1 In order to bring therapeutic substances into the body in such a way that they unfold their full effect in the right place without being absorbed or filtered by the liver, they require a specific chemical 'packaging', a so called 'formulation'. The formulation is one aspect of 'delivery', which also includes the ways in which medicines can be administered, such as orally or intramuscularly.

2 The institute was founded in an early governmental attempt to foster the application of biotechnology as a production method for organic compounds in the industry (see Adleberger 1999). Today it belongs to the Helmholtz Community. Like the Max Planck Society, this organization is a non-university, public research body. In contrast to Max Planck Institutes, Helmholtz Institutes do not exclusively focus on ground-breaking basic research, but on the systematic (one might say bureaucratic) integration of basic and applied research. Helmholtz Institutes are organized around issues (such as cancer research, infections research, environmental research), are interdisciplinary and tend to be very large. They are institutionally closely connected to federal governmental science policy and strategy formulation. Of all elements of the German research system, the biomedically oriented Helmholtz Institutes bear the closest resemblance to the American National Institutes of Health (NIH) system.

3 The same alteration was applied to morphine to create heroin.

4 The shift towards markets and viable business models

After their initial inception, to become successful, all innovations need to undergo substantial transformations. New ideas need to be materialized in a way that makes them work outside the lab and outside the still experimental assemblies realized in biotech start-ups. Furthermore, a target market and a business model need to be chosen and all efforts need to be concentrated on this one entry point into a market. Just as investors play a key role in seed financing innovations and putting them on track, they also are important players in the realization of such a 'market turn'. In this chapter I wish to discuss how biotechnology innovations arrive at their final business model and target market, which forms of agency and relational work are required for this, and what role investors have to play. Crucially, I do not see investors as the sole initiators of this change. As in early stages, they interact with other actors and forms of agency. My attention is on these interactions, on the succession of investors over time and on the interactions between different types of investors, each with a specific rationale and mode of knowledge valuation. First, the cases studied here will be categorized into business models and discussed within these categories. Then, I will draw general conclusions about key dynamics of relational work and the role of investors.

4.1 Pharmaceutical product development

In the early years of Germany's biotechnology industry, policy makers and entrepreneurs hoped to create a new kind of pharmaceutical company based on biotechnological expertise. The concept of the fully integrated pharmaceutical company (FIPCo), or fully integrated pharmaceutical company was both an aim and a benchmark against which innovative developments were measured. The stock market was considered an ideal source of financing for such companies, after a venture capital (VC) funded growth period. The 'Neuer Markt' technology stock exchange, established in 1997, was to serve this purpose. The role models for this envisioned development were American companies like Amgen and Genentec who, founded in the 1970s, had grown into important pharmaceutical players in the 1980s. These high hopes were thwarted by both the burst of the dotcom bubble in the capital market (and again during the financial crisis of 2009) and have never fully returned since.

> [...] The first aim is always to create a blockbuster drug. But that is not so simple anymore. Therefore, our programme is now called 'personalized medicine and targeted therapy', because you can hardly sell blockbusters anymore these days, since they all have side effects. And there are hardly any drugs left that fit all patients who are afflicted with a particular condition. One way to respond to this is to work on the formulation. The other is to try to use one substance for different indications. For example, a drug has been developed for colon cancer, but in actual fact it is not a drug against colon cancer, but against a degeneration of a particular receptor on cells, which causes the cells to duplicate continuously. That can happen in the colon, in the lungs or in the brain. And it is the same receptor everywhere. (Interview 1–3)

Drug development is characterized as an endeavour with an unsurpassed level of complexity.

> This is unbelievably complex. I always compare this to a flight mission to the moon. All the things you have to consider to actually reach the target! (Interview 2–1)

As one respondent explained, 'everything that goes into the human body' is subject to a very strict regime of extensive testing and evaluation until regulatory authorities grant market admission. This applies to new drugs, but also to new crops. Since the development of individual, genetically modified crops was not represented by an innovation biography, the following paragraphs will focus on drug development. Although a number of new paradigms for drug development have emerged partly as a consequence of the biotechnology revolution, such as 'translational medicine', the overall process is still structured by a regulatory framework which stems from the 1960s (Kaitin 2010).

A shared notion among biotechnology practitioners is that the testing and evaluation regime for drugs on the one hand can uncover scientific deficiencies in a new product ('scientific risk'), but on the other hand, can also impose its own kind of risk, which is rooted in key actors' uncertain willingness to accept the evidence and appreciate the value of a new product ('regulatory risk'). 'Risk' in this case refers to the product developer's chances of establishing a product in the market.

One of the preconditions for entering into standardized testing procedures is to have a sufficient supply of the respective substance in the required quality – and to have it at the right time in the right place. Production of pharmaceutical substances is regulated by GMPs or good manufacturing practices, one requirement of which is to have autoclavable[1] production machines. The high specificity of the task is met either by dedicated contract manufacturing organizations (CMO) or by production sites in pharmaceutical corporations. These sites need to be integrated into a multi-site network once biotechnology companies enter the formal stages of drug development. Accordingly, the biotech companies' original sites are partially transformed into hubs for logistical organization and the management of production relations. This includes, for example, the build-up of quality control capabilities for external products and services. Respondents provided different accounts of the distance sensitivity of production relations. One determining and very prominent factor in biopharmaceutical production is the stability and sensitivity of the respective substance.

> If a lorry carrying the drug were to have a mishap and defrost, that would be a disaster. (Interview 5–1)

After a potential new drug is identified in the lab and its underlying molecular interaction mechanism is explored, a phase of preclinical testing sets in. The preeminent purpose of this testing is to determine the potential toxicity of a substance and its ideal dosage before it goes into human bodies for further testing. For preclinical trials, a set of good laboratory practices (GLP) and documentation standards are in place, adherence to which is a precondition for admission to clinical trials. Preclinical tests are conducted in-vitro (in the Petri dish) and in-vivo (in the living organism). For in-vivo testing, animal bodies ('animal models') are used.

The subsequent phase of clinical testing is subdivided into several stages (Kaitin 2010; Nolting and Mietzner 2010). In stage-1 clinical trials, a new drug is tested in a small number of healthy human bodies to determine potential dangers and side effects. In stage-2 trials, a drug is tested on a small number of patients with the disease addressed to observe its effectiveness in a small group and to further specify the appropriate dosages. In stage-3 and -4 clinical trials, the effectiveness of a drug is tested on large numbers of patients so as to achieve statistically significant results. Hence, stage-1 trials are fundamentally different from later stages. In them, the subject is the human body 'as such', whereas later stages specifically address the patient population.

In the following stage 2, the drug is tested on patients with a specific medical disorder. At this point, product developers need to formally commit to a specific treatment regime to be followed throughout the trials. This includes the envisioned population of patients, a specific medical condition and indication as well as a specific form of delivery of the drug (such as pills or nasal spray). Until stage 2, a drug can have several potential applications. With the beginning of stage 2, one application is chosen and all other potential usages are cut off. This cut-off is not only a temporary prioritization, but a substantial backlash for future alternative uses, because the patent duration for the underlying inventions is limited. Picking up lose threads later may be hindered by lacking patent protection. Starting with stage 2, increasingly large numbers of patients are enrolled into clinical trials, until in the end, a statistically significant result, usually based on several thousand tested individuals, is reached.

Clinical trials are in most cases multi-local. They can be conducted at university hospitals, regular hospitals or specialized clinical trial facilities. The need to enrol a very large number of test subjects, and to create a representative sample of the patient population requires clinical trials to be organized on an international scale. The administration of such large-scale trials – managing the circulation of myriad documents and substances, but also creating a coherent picture of test person enrolment and enforcing methodological standards across multiple locations – is conducted by specialized contract or clinical research organizations (CRO). In the last roughly 20 years, numerous such specialized service providers have emerged, along with the formal occupational profile of the 'clinical trials manager'. Clinical trials are the most international activity encountered in the case studies. They span not only Europe and North America, but all continents.

Conducting a clinical trial is thus a specific practice with marked differences to both medical practice and basic medical research. The most pronounced of these differences is the absence and illegitimacy of any form of improvisation, spontaneous search or crossing of boundaries. It is fair to characterize clinical trials in the existent framework as a zone of forbidden serendipity and surprise (Cooper 2012). An example, which does not belong to one of the cases but was referred to by respondents, can illustrate this restriction, which produces its very own kind of risk and failure. A Regensburg-based biotechnology company conducted a clinical trial for a new drug to treat *residual postoperative* brain tumours. At the beginning of the development, tumour residues after brain surgery were a substantial problem. Attempting to treat whole tumours instead did not appear appropriate. However, during the development period, but independently from this particular development, surgical instruments improved drastically, leaving fewer and fewer tumour residues. As a consequence, not enough test persons with tumour residues could be enrolled as test persons during later stages of the trial. Redefining the objective to initial tumour treatment was not possible within the framework of one trial. Hence, the trial failed.

As with animal testing, different possibilities exist to deploy 'test balloons' to minimize the failure risk of clinical trials. Some of these have immediate consequences for the choice of product or indication which is advanced to clinical development.

The stated overall costs of preclinical and clinical testing differ drastically depending on the actor who provides the statement. In the case studies, figures from €50 million to €1.5 billion were encountered. A realistic estimate appears to be in the €100 million plus range. One respondent explained that costs above €1 billion, as they are often stated by the pharmaceutical industry, are to be understood as average costs per successful development, which includes failed trials. The chances of failure across the entire development timeline are considered extremely high, in excess of three-quarters of all developments. To small biotechnology companies, who are often funded by VC, the high costs and low chances of success render biopharmaceutical drug development an all but impossible undertaking.

Large pharmaceutical corporations on the other hand have the financial resources to conduct clinical trials and the likewise extremely costly market launches. In addition to the financial burden (and risk), the question of transparency of clinical trials is a relevant one. For small biotech companies, clinical trial entries, milestones and endings are 'a big thing'. They are key issues in investor communication and subject to vivid conversations in peer circles. Reports can be found in the trade press. There is, however, no general legal obligation to publish clinical trial entries, endings or results. Therefore, large corporations have a degree of leeway regarding the question of which trials (and results) are presented in an admission procedure. Surprisingly, there is also, as yet, no general procedural standardization of clinical trials across the medical community, which would render admission trials comparable to independent clinical research trials at university hospitals (for example, by providing a mandatory template for the reporting of side effects).

The successful completion of a clinical trial is a precondition of market admission by the central regulating authorities – the Food and Drugs Administration (FDA) in the United States and the European Medicines Agency (EMA) in the European Union. But in addition to the question of an individual trial's validity, the unresolved issue of publication raises the question of how certainty can be created in regulatory processes in the drug market, and which role is played by transparency and contestation. These questions are addressed by McGoey (2012). She argues that instead of creating shared knowledge, the key strategic activity of actors in drug market admission processes is to *claim non-knowledge and to battle each other for non-knowledge*. 'Strategic ignorance' is the consciously created absence of adversarial knowledges which could challenge the proposed claims of truth (such as knowledge about side effects). In this light, the (economic) ability not to publish clinical trial results produces a severe power imbalance between drug developing biotechnology companies and large pharmaceutical companies. Put very simply, having lots of money and a huge legal department buys the ignorance needed to succeed in an admission process.

Market admission is issued for the ex-ante specified indication and form of delivery. In addition to a successful trial, a therapeutic advantage of a new or improved drug vis-à-vis existing treatments needs to be demonstrated, such as fewer side effects, easier usage or better effectiveness. Generally, drug developing companies throughout the case studies strived for market admission in both the EU und the US with admissions for other countries ranging lower in priorities or later in the timeline.

Although being of comparable population size, both territories differ with regard to market attractiveness. Uniformly, respondents characterized the American market as exceptionally attractive for four reasons: its size, its purchasing power, its institutional homogeneity and the private health insurance system. US health insurance companies are praised for their readiness to reimburse the costs for new drugs soon after their market launch.

> Until now it worked like this: if you had FDA clearance for a drug, it was automatically reimbursed; because, if you kept it from the patient, especially in the USA, liability issues could arise. [Imagine] there is a new active ingredient which could have saved the patient, but the insurance company denied it. Then the relatives go to court when he is dead and say, 'Compensation. [*We demand*] one billion'. This is uncapped after all. (Interview 4–1)

Europe, on the other hand, is characterized as an institutionally hetero-geneous territory. Germany's public health insurance system as well as its governance relations with the legislative and the government in particular, are characterized as sluggish in making reimbursement decisions and as averse to the high prices of newly introduced drugs. Especially the requirement for new drugs, as viewed by respondents, to be cost-effective in the short term, even compared to generics (i.e. drugs which can be marketed cheaply because patent protection has run out), is considered adversarial to innovation.[2] To drug developers, a market admission in the United States is therefore the most attractive achievement. It represents, as it were, the crown jewels of biotechnology.

There are, however, two alterations or openings in the admission and market entry framework were encountered which proved highly significant to the pathways of biopharmaceutical product development: the institutional status of the 'orphan drug' (Mazzucato 2013) and new regulatory approaches to generics.

Given the high cost of pharmaceutical product development, medical con-ditions from which only a very small number of patients suffered did not represent attractive markets. Such 'orphan diseases' thus remained under-represented in drug development efforts. Legal changes in both Europe and the United States addressed this deficiency and an institutional status of 'orphan drugs' was created.[3] Orphan drugs can be marketed exclusively for ten years (EU) or seven years (US) after market entry regardless of the remaining patent duration. This period is relevant because the long duration of pharmaceutical product development usually takes up large parts of the duration of patent protection, thus forcing pharmaceutical corporations to realize high profits in a short time. This creates a bias for mass markets and quasi-monopolistic positions. The guaranteed time span of exclusive exploi-tation provides an opportunity to make a new product profitable in a small market. In addition, admission procedures and trial requirements are simpli-fied if a product in development is granted orphan drug status. Developers of orphan drugs, for example, cannot hope to enrol the usual number of test persons for clinical trials. Regulatory authorities also provide counsel and practical support to developers of orphan drugs. As a result, a distinct market for orphan drugs emerged when pharmaceutical corporations began specializing in them. To small biotechnology companies, the reduced uncertainty and quicker path to a market is a reason to select an orphan drug strategy. This choice was made in case 2, SYNTHESIS and case 5, AUTOIMMUNE.

The market for generics or 'imitator products' is widely considered to be a non-innovative field. Recently however, this dichotomous image of innovative and non-innovative drugs has received a re-evaluation. The reason is that the existing admissions regime, which implicitly favours patent-protected drugs, inhibits innovative re-combinations of established products if their patents have run out. Increasingly, drugs which were licensed for one medical condition are, in many cases accidentally, discovered to be effective against other conditions – often dependent on patients' genetic predisposition ('personalized medicine'). For example, in 2015 the diabetes drug 'liraglutide' was found to reverse effects of Alzheimer's disease in mice.[4] Publicly funded clinical trials were initiated. While, on the one hand, drugs which already have a market admission require no further testing for toxicity, the short remaining period or complete lack of patent protection reduced companies' willingness to invest large sums in effectiveness trials, thus necessitating public or civil society-based funding. While this fundamental problem remains unsolved, the admission authorities in both the EU and the US decided in 2012 to simplify market admissions for incremental improvements on drugs which have become generic. In case 8, NEURON this became relevant.

As seen in Chapter 3, generics as a business model can create the revenue to be used for highly innovative biotechnological venture funding (as in the case of the SOUTHWEST FAMILY VC office). Throughout the cases, several further incidences occurred in which the generics business model was brought in conjunction with biotechnology investment. Generics represent a field with substantial influence[5] of German companies such as Stada, Hexal and Ratiopharm. Ratiopharm was spun out from a traditional mid-sized pharmaceutical company located in the southwest of Germany in the 1970s. Thus, selling generics can be understood as a possibility to redeploy pharmaceutical production capabilities without having to bear the burden of biopharmaceutical product development. While generics are a safe way to create revenue, statements from interview partners indicate that entrepreneurs in the field, especially in family-owned companies, feel the need to create an added value rather than 'just copying' other companies' developments.

One entrepreneur related, in an anecdotal fashion, the history of his enculturation into biotechnology entrepreneurialism. Having spent his previous career as a pharmaceutical marketing executive, he was hired by the leader of a family-owned German generics company. The company leader intended to transform his generics company into a research-oriented pharmaceutical company with a strategic focus on one therapeutic field. The acquisition of one or several central European biotechnology start-ups was the envisioned strategy to achieve this goal. Consequently, the hired executive visited companies and spent time at the relevant trade events. This temporary legitimate peripheral participation (Lave and Wenger 1991) in the practice of biotech entrepreneurialism spawned his own wish to engage with biotechnology in the future, which he did. However, the family entrepreneur dropped his biotechnology ambitions. Instead, he purchased 50 square kilometres of oak forest from a personally acquainted, financially troubled aristocrat. According to the executive's slightly sardonic account, this acquisition also qualified as a long-term investment in objects of intrinsic value.

The anecdotal character of this account aside, in terms of 'orders of worth' (Boltanski and Thévenot 2006; Stark 2009) the entrepreneurial and scientifically bold nature of biotechnology appears to exert an attraction to more traditional entrepreneurs, even if their own area of expertise would generally be considered very distant from it. Shifts in business focus can lead to an entry into the field of biotechnology (and therefore a redrawing of its boundary). But there is no determination, since the more general entrepreneurial value appreciation of *Mittelstand*-type family entrepreneurs (creating a permanent value, doing something intrinsically valuable) can also apply to other things. In case 6 STEM CELL, a successful entrepreneur in the field of machine construction acted as a private investor in a newly founded biotechnology company, based on this rather 'irrational' value appreciation and without experience in the field. Like the case's other private investors, he remained committed to the company, although the investment could not be considered a success in financial terms.

In case 5 AUTOIMMUNE, a start-up company had been created in the mid-2000s by an immunologist who had worked in a Max Planck Institute (MPI) in Munich. The company had received mentorship and institutional support from Max Planck Society. An Italian pharmaceuticals company, acting through their corporate discovery VC fund, had placed the first major investment out of strategic interest: They sought to develop a range of new autoimmune drugs. However, the fund itself was managed in a more VC-typical way: aiming for growth in the portfolio companies' resale value, thus adopting a financial logic rather than a pure strategic (i.e. oriented towards the mother company's product development goals) logic. This would become relevant during the innovations' development trajectory.

After several years the investors grew unhappy with the management performance of the original founder – a dynamic which would show repeatedly across the studied cases. They looked for a new chief executive officer (CEO). This new CEO turned out to be the very industry representative who earlier had collaborated with the Max Planck Institute's leader on initiating the science-industry collaboration which inadvertently provided the opportunity for the new company's founding. While there is no compelling causal link between the two events, personal acquaintance with both people and technology – as in case 3 GENE FUNCTION – appear to be relevant factors for a successful match. After a long career combining academia, industry and repeated start-up entrepreneurship, the CEO accepted the call as a last assignment before retiring. He moved home to Munich for this purpose.

His aim was twofold: Gearing the company towards a promising development strategy and finding new investors which would allow the company to take at least one drug into clinical development. To this end the company's internal structure had to be streamlined around a few key projects. Processes became more formalized to comply with regulatory requirements. The new leader took an empathic approach to convincing the staff of the shift:

> If you talk to people, if you communicate what the common goal is, then scientists too will understand where the limitations and boundaries are. Which priorities to set, where to cut? [...] It was the first time in my career that the group itself realized the shift of projects, in group activity. Instead of saying, 'your project is cancelled', people can say, 'I get to work on a project with a high priority'. (Interview 5–1)

Yet at the time in 2008, in the midst of another financial crisis, neither an industrial, nor a VC investor was ready to invest in the early stage product development. The CEO followed the usual procedures (such as pitching on international investor conferences), but found no interest. At a local event in Munich, he became aware of RETAIL VC (see Chapter 2). The company matched their investment approach: As it was developing a biopharmaceutical product in an early stage, the company was considered a normal investment by RETAIL VC. Thus, this then very young fund became the first investor in a second wave of substantially larger investments. Again, there is no causal necessity for a proximity-based relationship (this time in the literal sense), yet co-presence in Munich helped in creating a match. Crucially, the company at this point had an experienced management team – a precondition for an investment by RETAIL VC. RETAIL VC also helped to enrol SOUTHWEST FAMILY VC as well as a Swiss VC fund as investors. Expanding the investor consortium with VC required the original investors' acceptance and openness. The investee company subsequently brought its lead product, an orphan drug, into clinical development.

In terms of product development, a variety of autoimmune disorders with large patient populations such as lupus, rheumatism or multiple scleroses were available for selection, since the underlying invention was related to misdirected immune responses in general. The leadership chose to enter clinical development with a treatment for idiopathic thrombocytopenic purpura (ITP) instead. In this disease, the immune system destroys the blood platelets necessary for wound closure. Although having only a very small patient population, the condition – in addition to being an orphan disease – has the 'advantage' of a very early 'clinical end point': The success of the treatment can be determined after days, since all that is required to be measured is a recovery of platelet counts in the blood. Observing the recovery of an inflamed joint on the other hand, as in the case of rheumatism, requires months. Identifying a trial as failed early on reduces costs and thus increases the chances to get a second shot – a rare occasion during drug development, which is usually a 'one-shot operation' (Rittel and Weber 1973).

The selection of a drug for an orphan disease as the first – and potentially only – product locked the company into an orphan drug and niche-market-oriented strategy. In theory, the ITP product could act as a test case for following, more mass-market-oriented products. This attempt at maintaining technological coherence is however endangered by the limited duration of patent protection. In addition, from a market strategy perspective, the difference between a mass-market- and a niche-market-oriented product creates a potential breaking point in a company: Biopharmaceutical companies which persist as organizations (rather than being temporary projects) tend to be specialized on either mass or niche markets. In 2013, the lead product was in a late stage of clinical development and showed very promising results (i.e. it was close to the market). It was complemented by an additional drug development for a more common autoimmune disorder; this more mass-market-oriented development was prioritized by investors and likewise entered a late stage of clinical development.

During the investment period, RETAIL VC invested in four rounds and enrolled further investors, including an American venture fund. Over the course of seven years (five since the entry of RETAIL VC) around €52 million (€48 million) were invested in 5 rounds. From the perspective of the investors, the combination of an orphan drug development and a mass-market drug development posed a challenge regarding the envisioned exit. Pharmaceutical corporations typically specialize in one of the two. It was therefore probable that the company would be split up and the product developments would be sold separately (as in case 2, SYNTHESIS).

However, in March of 2015 an American pharmaceutical corporation acquired the investee company wholly. This strategic investor had a specialization in immune deficiencies and announced that all development operations across a range of envisioned products in Munich would be continued. Hence, financial value and technological interdependency were brought in congruence, and the founders' original idea to establish a multi-product company based on technological interrelatedness actually succeeded. A precondition was the presence of investors (SOUTHWEST FAMILY VC and RETAIL VC) who had both the necessary size and a sufficiently wide investment time horizon to fund clinical trials and to await the fitting exit opportunity.

In case 7, CANCER IMMUNE, a start-up had been founded out of Hamburg Eppendorf university hospital. Its aim was to apply a new gene vector technology to cancer therapy. The field of liver cancer had been chosen, because the technology's effectiveness could most easily be proven in this type of cancer. In this case, a traditional, medium-sized pharmaceutical company acted as a strategic investor by striking a cooperation agreement with the technology developing start-up company. The objective of this – at the time of data collection still ongoing – partnership was to take one drug out of an array of possible applications into clinical development. This incident is exceptional, because the traditional *Mittelstand* type pharmaceutical businesses are usually considered to be excluded from the new form of biopharmaceutical product development due to insufficient size, managerial capabilities and financial resources. Unfortunately, no interview could be conducted with a representative of said company. Therefore, aside from stating that a novel linkage between hitherto unconnected social orders was in fact created, little can be said about the investor's action logic.

However, the details of the investor's integration into the innovation's unfolding network offer insights into the time-spatial organization of such a new association. The investment relationship was brokered by the private investor or 'business angel', who also acted as mentor and strategic advisor, and who had encouraged the company founding in the first place. The investor, who had previously been a company founder and a consultant to diagnostic and pharmaceutical corporations, was personally acquainted with representatives of the investing company as a consequence of earlier collaboration experiences. Personal acquaintance and previous interaction experience, as in other cases, were seen as a helpful and possibly indispensable influence. This was particularly true with regard to the market strategy and current position of the potential strategic investor; as in other cases, the investing pharmaceutical company was in a position in which an existing product was about to become generic in the foreseeable future and a replacement had to be envisioned. Hence, defending a market position was the motivation for engaging with a biotech start-up.

The private investor stressed that knowledge of such a sensitive strategic positioning process and access to the negotiating table required pre-existing trust. While she argued that any international pharmaceutical company could have acted as an investor, in actual fact, the investing company was located in very close spatial proximity of the biotech start-up (both in and around Hamburg). Furthermore, as in other cases, the increasing importance of market strategy which came with the investor's enrolment led to a shift in the product development focus. Initially, both the company founder and the private investor had a) put emphasis on the potentiality, i.e. the wide applicability of their technology in oncology as well as other fields, and b) prioritized a type of cancer where the technology's performance could be demonstrated most effectively. Under the influence of the new investor, the priorities were shifted in favour of a product development fitting the investor's market position.

As in this case, some seasoned entrepreneurs and private investors appear to pursue the approach of circumventing VC entirely and partnering up founders and strategic investors directly. One respondent presented two rationales for this approach. Firstly, strategic investors in the form of pharmaceutical companies possess the managerial and technical capabilities to develop pharmaceutical products in-house. Hence, there is no need for building such capacities in a biotechnology company or for sourcing them at high costs from specialized service providers. And secondly, VC investors have unrealistic expectations of financial returns, which unnecessarily reduce the scope for substantial innovations. The private investor, who had brokered the industry investment in case 7, explained:

Gosh, I invest sums around €100,000, no more. And I would say [...] one million is the upper limit for a private investor. Because I think, What's this about? There are companies which have everything, you know. They have departments for admissions; they have whole departments for production, for quality control. They have all that on standby. Why should I not do it along with them? This is something I saw in my own company; it pays off to have cooperation very early on. And it's better not to have the big money dangling in front of you, because when I have the big money in my mind's eye, that's a chance of a hundred to one or five hundred to one. [...] There are people from the industry, from around here. There is our former minister-president Späth and people like him, they have tens of millions they can invest. And I try to tell them that they have a very short-sighted perspective. It's always about making the big profit quickly. But this way we choke these promising little companies. It's their death. (Interview 7–2)

Additionally, the cost saving argument in favour of strategic investors also serves to rebut one of the main legitimizations for VC in biotechnology: the – alleged – high cost of biopharmaceutical product development.

In this case, High-Tech Gründerfonds (HTGF) (see Chapter 2) invested as a public VC investor after the private investor had taken the lead. The fund was approached by the company founder, but via another VC investment company who frequently acted as broker or 'scout' as a part of its positioning strategy. The investment decision was made after HTGF found that the company's technological approach had the potential to fundamentally improve cancer therapies. Hence, technological innovativeness and potential were the investment criteria. The founder considered HTGF's investment as essential to his efforts to establish company operations. However, the fund was not perceived to provide a workable strategic perspective.

[...] Actually, I had hoped that High-Tech Gründerfonds would support the companies more, like it promises to do; in a consulting manner. Sure, they establish contacts; there is an event where you can meet investors, institutional investors [*i.e. venture capitalists, author's note*]. But if these institutional investors do not invest in therapeutic product development, the whole thing is futile. (Interview 7–2)

Throughout the observed innovation process, the first investor (also mentor and strategic advisor) remained highly influential. No marked leadership change could be observed. However, a former executive in the pharmaceutical industry had been part of the founding team in a very early stage but left the company in conflict about the company's scale of operations and business strategy. A preference for 'staying small' and searching for the right opportunity with an industrial investor had, according to respondents, prevailed over a more up-scaled and cot-intensive idea of business development.

In case 8 NEURON, similar dynamics played out. A company had been set up in Mainz in 2005, whose purpose was to develop a chemically altered plant-based compound into a new Alzheimer's drug. The invention had been made during research on nicotine receptors in the brain. After building an initial investor consortium consisting of HTGF, a regional public VC fund as well as KfW (Reconstruction Loan Corporation) as a matching investor, no further investors had been found. The development had stalled, the company existed solely as a name. Around 2010, two strategic investors occurred, one occupying the additional position of a 'build-up partner' (i.e. providing expertise and technological capabilities for large-scale development and production, Ibert and Müller 2015) and one acting as a 'pioneer user' (ibid.). The earlier was, once again, a more traditional, *Mittelstand*-type company, in this case a South German production company in the tobacco industry. The substance on which the novel drug developed in this case is chemically based, similar to nicotine and can be chemically synthesized or extracted from plants – including tobacco plants. While the founding team did in fact possess a patent for chemical synthesis of the substance, advances in extraction technology and the discovery of large-scale natural reservoirs in plants in Asia made extraction the more efficient choice for production. Confronted with a diminishing market for tobacco products due to increasingly strict regulation of smoking and cigarette marketing, the tobacco company sought opportunities for redeploying its capabilities.

In this position, the company's leadership made the decision to engage with the drug development pursued in this case. This included investments both in the drug developing biotechnology company and in its own production capabilities, which needed to be reconfigured for extraction as well as to meet pharmaceutical production standards (GMP). The latter activity also included investments in China, as extraction technologies had evolved there. The decisive investment rationale was that, in case of a successful market launch, the investing company would become the exclusive contract manufacturer for the drug. Hence, a highly innovative material association was created, which was driven by the motivation to enter a market and build a new market position. The shift from tobacco processing to extraction and drug production could be understood as an application of 'related variety' (Frenken, van Oort and Verburg 2007) or relational proximity (Boschma 2005). But it should be clear from the circumstances of the engagement that the specific action situation and its interpretation (diminishing market, chances in the pharmaceutical market) were behind this association, rather than an objective and inherently stable property of relatedness.

In the same case, a Canadian generics company acted as the second strategic investor and pioneer user. This company pursued a strategy of entering the market for *patent-protected* (i.e. non-generic) drugs addressing neurodegenerative diseases. As described above, regulatory authorities in both the US and the EU had moved towards a prioritization of incremental improvements of generic drugs, thus making the boundary between generic and patented drugs more permeable. This regulatory change provided an opportunity for a generics company to enter the 'real' pharmaceuticals market using an incremental expansion approach. Interestingly the usual dynamic of shifting product development foci – from general potentiality to one isolated application – was reversed in this case. The company founder had envisioned a market admission for an Alzheimer's therapy, Alzheimer's disease being the most promising market in the field of neurodegenerative diseases. The investing company, however, strived for a comprehensive positioning in the field of neurodegenerative disease therapies and entered clinical development for a whole range of indications in the field.

In sum, case 8, NEURON, shows a very unique pattern of associations between 'innovative' and 'traditional' (in the prevailing perception) things. Very advanced research on neurobiological mechanisms (specific receptors in the central nervous system and their importance for cell survival) led to insights into the effects of a long-known, naturally occurring substance. A likewise long-known chemical modification was found to make the substance more useful as a drug. Advances in production technology allowed a traditional manufacturing company to jump the drug development train. Finally, gradual changes in the regulatory framework allowed a generics company to build a radically new market position.

In case 8 NEURON, the entire development process was driven forward by one key individual, while in most other cases a variety of individuals were involved in organizing the shift from a more science-driven to a more commercial enterprise. As discussed in Chapter 3, he perceived the long-lasting lack of follow-up financing (after the first public and semi-public round) as a consequence of VC companies' dependence on trade sales to the pharmaceutical industry (which pursued a scientific trajectory different from his own). Consequently, his preference was to wait and search for the right opportunity outside the established VC-pharma link. This orientation put him at odds with his investors.

> As it happened our investors [...] had the opinion that we probably weren't very good at marketing [...] and for the money we got from them they implanted a very expensive so-called business developer [in the company], that is, people who have a good reputation and get incredible lots of money. He [...] got €2,000 per day plus expenses and travelled around to find us a big company as a licensee, and of course this doesn't work. We knew that. Because actually we had our contacts with big industry and we knew why they didn't want it. And for these nine months he blew about half a million. (Interview 6–1)

4.2 Diagnostic product development

Diagnostic product development constitutes a path to the market which is considerably less complex, risky and expensive than pharmaceutical product development. For some technologies, which can potentially be used for both, the diagnostics path can be an alternative to the drug development path.

Throughout the studied cases, respondents made frequent reference to the specificities of diagnostic product development vis-à-vis drug development. In these accounts, the field of diagnostics is likened to the field of medical technology. Both are said to share key characteristics, such as their risk of failure, their development costs, their development timelines and the complexity of admission procedures – all of which are lower or shorter than in the case of drug development.

In case 4, BIOMARKER, a methylation company with an unclear business case had been set up and VC-funded in Berlin, while a likewise methylation-based company focusing on blood-tests for early stage cancer diagnostics in Seattle had not been able to attract VC. In this rather improbable situation, the Berlin-based start-up got into the position to initiate a merger. The lead-investor in the Berlin company was a Munich-based VC fund. The VC investment manager succeeded in bringing in other VC firms in Europe (notably the UK) and finally also the United States as co-investors. Thus, a formidable transatlantic investment consortium was created. Using their position on the company's board, the investors pressed for a less technology-oriented and a more business-oriented course and thus prioritize the only business model, which appeared realistic at the time: diagnostic product development.

At a methylation conference in Germany, the leading Berlin founder and representatives of the American diagnostics company met coincidentally. After further talks, and with the backing of the investor consortium, both sides agreed on a merger or rather, given the financial power relations, a take-over. Henceforth, the company would be run as a two-site enterprise with the Berlin plant housing the senior management team. The latter included members of the original Seattle founding team. Both sides were not only different in terms of age and experience (young, technology-driven founders vs. experienced clinical experts), but other differences became apparent, which were attributed both to national working cultures and different market structures. This point is stressed, because in the field of diagnostics, national 'cultural' differences appear to be a more relevant subject in product development activities than in the case of drug development.

Initially, both sides participated equally in research and product development activities. Coordination between the sites required a number of adjusted working routines. One was the synchronization of working hours to deal with the nine-hour time difference. Employees in Berlin tended to work late whereas employees in Seattle worked early to maximize the temporal overlap and to accommodate video or phone conferences. Another measure was a staff exchange program. Individuals with key responsibilities in the development process experienced life in both localities and work in both sites.

While having received substantial funds from a transatlantic VC consortium, the company lacked the means to develop a product and launch it in the market on its own. Therefore, a product development partnership with a large Swiss diagnostics company was initiated – encouraged and supported by the investors. In the underlying division of labour, the start-up company was to further develop their key technology, which would then be integrated into test kits for diagnostic laboratories by the Swiss corporation. The latter funded the collaboration and provided expertise and infrastructures. The key challenge lay in producing proof that the start-up's basic technology could feasibly be run, producing reliable results, in high-throughput and with large sample numbers.

The Swiss diagnostics company invested substantial funds (eight-digit figures, in euros) to have a lead diagnostic product for colon cancer developed, which would, in the case of success, be followed by products for other types of cancer. Hence, the company acted as a pioneer user. Since it also contributed skilled statistics experts and high-performance gene sequencing machines, this strategic relation also had qualities of a build-up partnership (Ibert and Müller 2015). Both aspects taken together indicate that, rather than a customer–supplier relationship, this could be qualified as an investment-relationship. Both the technical aid and the funds (which required no payback) produced a substantial leap forward to the Berlin-Seattle start-up company. This leap was of additional value because the user company did not claim any intellectual property rights (IPR). In practical terms, the relation materialized itself first and foremost in circulating samples, data and documents and to a lesser degree in site visits. This specialized and complementary division of labour was possible because the start-up company had already achieved the necessary degree of professionalism to organize knowledge work across physical as well as relational distance.

As most respondents recall, the relation was experienced as highly productive by all direct participants on the technical level. In retrospect, the relation offered both challenges and opportunities in addition to the linear progress in product development. While it was still active, the start-up company's management team, with the support of the investor consortium, brought the company public successfully. Rather than realizing an exit, the investors remained on board as stockholders. At the time, the mid-2000s, a successful initial public offering (IPO) of a biotechnology company was an unlikely and widely considered impossible event in Germany. At this time, the exclusive strategic partnership with a world-leading diagnostics company was a favourable and decisive argument to buyers on the stock market. This mode of value justification on the stock market differs from what Lazonick and Tulum (2011) say about the relation between stock market and strategic investment (in the United States). According to them, the speculative, risk-affine American stock market acts as a secondary market for strategic investors (pharmaceutical corporations) who want to get rid of unsatisfactory investee firms. In this case, stock market performance (on the German stock market) was dependent on the interventions of strategic investors.

However, for reasons not entirely verifiable, the relation with the Swiss diagnostics corporation was terminated abruptly. Most respondents attributed the end of the relationship to a shift in corporate strategy on the user side, which rendered the project irrelevant – despite the substantial progress it made in scientific and technical terms. A minority view attributed the sudden rift to the start-up founder's side. When the partnership failed, the young transatlantic methylation/diagnostics company was required to develop its own market entry strategy. The exclusivity and high degree of specialization of the previous relationship, however, was experienced as a massive downside after its ending. Linkages to the market and to end-users now had to be built from scratch. The absence of any financial or IP-related obligations, on the other hand, made the necessary reorientation more feasible. The user company had essentially financed a leap forward, towards a technical proof of feasibility, without getting any marketable result aside from a deeper understanding of the technology.

At this point, the young firm was left with the challenge to continue product development on its own. To become a product company, it had to undergo dramatic changes in terms of working culture, leadership and the spatial organization of knowledge work. In this phase, the original lead founder left the company, dissatisfied with on-going developments, and the search for a new CEO ensued. Existing inter-personal networks (on the side of investors as well as management) were used to approach an experienced executive from another international diagnostics corporation. In the same way as in other cases, he assessed the situation for its risks and potential rewards before moving to Berlin and taking the job.

> But this is of course a risky affair. Because you give up the company pension and the stock options and everything else you have in large a company like [pharmaceutical company]. And then you go full risk with a company like [*start-up*], because if we hadn't met the interest of investors, we wouldn't be here today. (Interview 4–3)

The new CEO lead the company's transformation from a technology-centric company to a product-centric-company:

> We had a big employee meeting. I said, 'Actually I have only one vision for [company name], and that is a little box. And this little box contains reagents which we produced. And these little boxes, they will stand in many molecular diagnostics laboratories, where the laboratories paid more than it cost us to make them'. And there was a sudden 'Ah! That is why'. We are not here to create the next methylation technology and even better marker discovery and to optimize one more technological aspect. We are here, at the end of the day, to bring products into the market. (Interview 4–3)

The original experimental and science-driven approach to the key technology was replaced by a more standardized, routinized and formalized approach, anticipating the requirements of regulation authorities and their admission procedures. Researchers employed in the company experienced this shift as a dramatic change. A substantial part of them left the company. New experts with specific training for the task were hired. As one executive explained:

> Try telling a laboratory researcher who used to work at a university, 'No, you will pipette it like this exactly and not in any different way, and of every slightest deviation you will make a note. You will make a note of every single step you take. And you will do it the same way over and over again'. Try saying this to a university researcher. He will rip your head off. [...] And, yes, building a quality management system was an important point. Very strong labour pains. Terrible for the people. All of a sudden, you have to observe SOPs, standard operating procedures, LABs, laboratory instructions, folders like this! [*Indicates size of folders with hands*] Everything has to be documented. (Interview 4–1)

As in the collaborative project which preceded the reorientation, colon cancer was chosen as the objective of the lead product development. The condition was chosen for several reasons: the size of its market – resulting from large ageing populations, its high incidence in rich countries with formal requirements for preventive medical check-ups – and its good treatability in the case of an early discovery which adds attractiveness to a diagnostic product. In addition, no blood test for colon cancer was available so far. As in the case of drug development, the US market was considered more attractive than the EU market, specifically Germany. The same reasons were given as in the case of drug development.

Europe and America differ strongly with regard to their market entry conditions for diagnostic products. As respondents explained, the American FDA requires clinical trials to prove the effectiveness of a diagnostic test. These trials do not, however, follow the same sequential hierarchy as drug admission trials. The reason is that in the case of a diagnostic product, no substances are brought into the human body. Instead, blood samples are taken and analysed. To obtain a US admission, an American clinical study was initiated. To this end the Seattle site, which became increasingly dominated and functionalized by the Berlin site, was transformed from a research location to an administrative centre. Laboratory spaces disappeared, office spaces expanded, the personnel composition changed. In this new sub-centre, the collaboration with numerous American clinical trial sites and university hospitals as well as correspondence with the FDA were orchestrated. In a large-scale clinical study, the effectiveness of the colon cancer test was established. In spite of the complexity and resource intensity of this enterprise, the diagnostic path toward the market is substantially easier and has fewer barriers than an entry into the pharmaceutical product market.

In Europe, a formal market admission for diagnostic products which already conveys proven effectiveness does not exist. Instead, product developers are required to certify their products themselves. Such a CE certification can be obtained by defining an aspired performance level and then testing the product against this individually defined reference. A certified performance level is formulated as a simple statistical relation: for X per cent of all individuals who actually have a medical disorder, the test will be positive. Of all individuals for whom the test is positive, Y per cent will have the disorder. The testing necessary to establish this relation is to be conducted according to Standard Operating Procedures (SOP) provided by the International Standardization Organization (ISO).

Compared to drug admission, this represents a very low market entry barrier. But as a consequence, there is also no striking argument in favour of the new product. Therefore, product developers need to gradually convince users of the validity and value of a product, thus negotiating their way into the market. Respective strategies of enrolment are based on multiple layers of representation. The users of diagnostic kits and hence customers of diagnostics companies are dedicated diagnostic laboratories. These laboratories will offer specific diagnostic tests if these are requested in relevant numbers by family doctors or general practitioners (GPs). GPs will recommend certain tests to their patients and request them from laboratories if they are convinced by medical specialists – in the case of colon cancer gastroenterologists and oncologists – that these tests perform better than their competitors.

Therefore, diagnostic product developers strive to enrol medical specialists, 'opinion leaders', as their allies. This involves, for example, being present at conferences of the respective communities. Such negotiations with gate-keepers, combined with the smaller importance of clinical trials, represent a commonality which diagnostic product development has with medical technology product development (e.g. prostheses). In both cases, the market entry is a more gradual, organic process. Revenues can be generated earlier on. A greater variety of partially collaborative and partially competitive relations is possible. Relations with users are more interactive and allow 'permanently beta' (Neff and Stark 2004) style learning cycles.

In case 4, BIOMARKER, interaction with customers – diagnostic laboratories – led to gradual improvements on the original product. At the same time, customers were taught to use, understand and appreciate the product (see also Jeannerat 2013).

> We send our technical support staff into the laboratories. And there they learn how this is done in practice. [...] then the customers start testing the product. [...] Before we launched our serial product, we introduced a research only version into the market. And that was a quick way to make the product available to the individual laboratory. And as I said, we trained the laboratories. And then they started testing and came back to us and said, silly example, 'Guys, could you ... we have these white bottles with reagents in them. [...] Could you perhaps mark the enzymes green and this red and this blue, because that would be much simpler for us. And also there are three bottles, why not just one? [...]' So, and what this means is that we got very specific hints about how we could improve

the product; and more from a practical perspective from the [user] laboratory, not with an academic perspective as in our own development. (Interview 4–3)

National working cultures were experienced to be very diverse. To respond to this diversity, a different approach in product design and accompanying interessement activities had to be chosen.

> Huge differences. Take an American and a German laboratory. In the German lab, you have highly trained staff. In the American lab they take on staff ... there is no big training or any such thing. They take people directly off the street. They only learn to press the right buttons on the machine. This means that the requirements for a product are completely different in the USA than they are in Germany. Then there are other cultural differences. The French always like to have things a little fancy. The Italians always want the newest stuff, never mind if it works, only if it's the newest. Germany is very conservative, very academically oriented. There you need tons of studies that show that it does exactly what it does and nothing else. Yes, there are huge differences. (Interview 4–3)

Likewise, marketing and distribution channels had to be developed for different regions. Germany, Switzerland and Austria were served by the core company sales team. For France, a separate but company employed sales force was established. In order to expand into further regions of the world, such as the Middle East, specialized sales intermediaries were hired as service providers.

To increase market penetration further, non-exclusive partnerships with four established international diagnostics companies were established. On the basis of a non-exclusive license, the four corporations used the newly developed diagnostic reagents. All five partners integrated them and marketed them as components in their own branded diagnostic kits, utilizing existing customer relations and exploiting existing market shares. In addition to individual brand names, all five partners collaborated in communicating the scientific denomination of the underlying diagnostic principle to opinion leaders. While diagnostic laboratories were thought to associate the new product with company-specific brand names and user interfaces, doctors were expected to use the scientific denomination to request the product. These collaborative activities were directed at creating a hitherto non-existent market sub-segment for blood-based colon cancer tests.

The number of four non-exclusive strategic partners was characterized as an industry specific optimum, based on a heuristic or rule of thumb which had been acquired through personal experience.

> Four. After 25 years in diagnostics, it tells me that this is a good number. [...] There is only a limited number of companies with a real-time PCR platform, and that is the technological platform on which our test runs.

[...] Not 50 or 100 players. [...] You want a number of partners that grants you market access. [Diagnostics company] has roughly a market share of 20% of kits placed in laboratories. [...] If you go up to roughly four players, you get access to the whole market. (Interview 4–3)

In drug development, even biotechnology companies which are considered successful do not distribute their own products on the market. Instead, they license their products to pharmaceutical corporations at a late development stage – or are bought by them. The reasons given are the high degree of centralization on the pharmaceutical market and the associated enormous costs of a market launch. Biotechnologists view pharmaceutical corporations as the natural actors in sales, marketing and distribution. As the case cited shows, in the diagnostics market, biotechnology companies have the opportunity to develop the competences which are required for market penetration, themselves or in collaboration. The original founding sites of biotechnology companies thus can transform, but retain their centrality in all respects. Differences in user practices, which are experienced as territorial-cultural differences, can be experienced by product developers in an immediate manner. A more diverse, decentralized and multi-voiced articulation of values and preferences takes place.

Around 2011, the company began marketing its first product, a blood-based cancer test, in the European and American markets. The company remained independent and established itself as a stock market-listed diagnostics product provider beginning portfolio diversification (two different cancer tests at the end of data collection). The original investors remained invested well into the company's maturity and long after the company went public. The temporariness of a VC fund usually would prevent such long-term commitments. However, without specifying, respondents hinted that the Munich-based VC fund, which acted as first investor, was special in this regard.

4.3 Hybrids and platforms

Of all biotechnological business models, through which an innovation process can pass or into which an innovation process can lead, the platform is likely to be the most elusive and underexplored. There appears to be a superficial certainty about platforms which is shared among participants in the biotechnology field. The platform is widely understood as a low-risk, service-oriented business model. In it, a particular technological competence is harnessed and stabilized in socio-material routines, which allow for a reproducible performance and thus a service product. Such services include, for instance, research and discovery, testing, synthesis and small-scale production.

The customers of platform biotechnology companies are large industrial corporations, typically in the pharmaceutical sector. Usually, the products of platform companies are research and development (R&D) services directed at product-development or product-upgrading efforts in pharmaceutical companies. Thus an R&D service market has emerged which is a major source of revenue to biotechnology companies, particularly in Germany. This market is considered more accessible to young start-up companies than the therapeutic or diagnostic end user markets. In a way, the relationship between 'big pharma' and the biotechnology sector – biotech as a dynamic field of evolutionary knowledge creation and pharma as a stable commodification structure – is consolidated and stabilized in the organizational form of the platform company.

Very often, the platform business model is framed as a low-risk alternative to a product-oriented strategy; hence, both models are treated in a dichotomous manner. However, platform companies are rather diverse. Furthermore, platform companies can participate in the creation of (monetary) value in pharmaceutical product development processes in different ways. One field expert explained that the relative positioning of biotechnology companies via pharmaceutical product development is of particular relevance for a company's likelihood to find customers. Given that pharmaceutical customers themselves find themselves in a volatile market, early-stage discovery services for new products are more difficult to market than later-stage product development services.

> How are the customers doing? We have of course many companies which offer early-stage research services. A company in Potsdam named [company name] offers molecules: molecules which the pharmaceutical industry than screens for their properties. Those who offer, 'I will do the clinical trials for your products', will not be scrapped. This is something you want seen through, you want to sell something. But in the case of early-stage services [customers] will say, 'Next year perhaps', or, 'on a smaller scale'. Especially with such services it depends: how are the customers doing? Are the customers undergoing mergers? In the beginning of 2009, we had the situation that of the top 20 pharmaceutical companies, six were undergoing a merger, i.e. 30% of potential customers [...]. The market volume was very small. [...] Globally, there are perhaps 5,000 companies in red biotechnology [...] and they buzz around the top 50 of the pharmaceutical industry on a daily basis. This means: 100 to 1 is the relation we are talking about. (Interview e–1)

This positioning of platform companies, as it were, at the low end of the food chain sheds doubt on the notion of platforms representing a low-risk business model. Preferences and perspectives of investors regarding platform companies were highly heterogeneous in the cases studied.

As networks of artefacts, practices and human and non-human activity, localized in particular sites, platforms are essentially extensions of the experimental work which is always necessary to stabilize innovative ideas. In all cases, innovative ideas required environments in which repetition and continuous adjustment were enabled materially – typically in the laboratories of public research facilities, but also start-up companies. Therefore, the technology platform as a socio-material arrangement and at the same time an organizational form is the 'natural' way to frame the early stage of a biotechnology innovation, at least as long as a dedicated product development framework is absent (although, as mentioned before, product development is the long-term aim most entrepreneurs have in mind). In fact, the cases in which exclusively product-oriented strategies were pursued by enthusiastic founders and their teams were the exception in the cases studied. In the following paragraphs, I will cite examples of platform companies some of which also engaged in their own product developments (i.e. hybrids). I will focus on the platform aspect in order to clarify what platform is or can be.

Sunder Rajan's (2006) characterization of biotechnology as being in part an information technology is helpful to understand platforms. On the one hand, organic materials are always the object of work. On the other hand, the creation of information, of statistic relations, tables and datasets, is always an important aim of the work. Databases are important intermediaries; computers are important non-human actors. Platforms as well as the knowledge practices pursued in them oscillate between the production of organic materials and the production of valid information. On the path towards one possible market, platform companies need to specify what the commodity is they offer and how it relates to product development for end-user markets, and to build appropriate socio-material networks across places. Especially in processes of up-scaling and automation, a once-chosen path is materially stabilized and thus path dependency is created.

In case 1 ENCAPSULATION an early strategic investment by an American pharmaceutical company had allowed a founder to assemble a team and a small-scale technological platform in his Munich start-up company. When the pharmaceutical company encountered severe difficulties with a previously launched product, all external engagements were ended. The enthusiastic founder lost his personal ally in the company and consequently all support. He and his team were left without a job for roughly six months. The team sought to identify and enrol new investors, but since it was the year 2002, shortly after the burst of the dotcom bubble, the established approach of appealing to investors' passion for the innovative technology remained fruitless. Finally, responding to an ad, the team got in contact with a British–French pharmaceutical corporation who were looking for an improved method to encapsulate one of their active ingredients (an established product). Presenting both their data and their equipment at the company's plant in Paris, they ensured the support of the company's head of research and one of his leading executives, who would become the team's main contact in their daily work.

In contrast to case 4, BIOMARKER, the earlier strategic investor in this case had retained IPRs (patents) – possibly in the hope to receive royalty payments from any future user or as a protection against potential competition. When the founding team entered negotiations with the new strategic investor, this pre-existing entanglement was an obstacle. It had to be removed in lengthy negotiations. The executive in the investing company, who performed the day-to-day interactions with the founding team and also led the patent negotiations, had, on other occasions, negotiated with the patent holding company in his previous position. He knew the individuals involved – a fact to which he partially attributed the success in easing the 'stranglehold'. In the new strategic relation, the start-up company kept all IPRs but agreed to never use them to create a product which would directly compete with one of the investing company's products.

For the time of the collaboration project, the aforementioned representative of the investing company would bring a more business-development-oriented logic into the working routines. From the perspective of the investing company, this relative lack of business acumen on the side of the founding team was perceived as an advantage: It translated into a high degree of openness, scientific curiosity and transparency as well as a collaborative attitude.

> You know, [*enthusiastic founder*] is not a businessman. He is a very good scientist, a very collaborative guy, interested in working with partners, has given a lot of the background knowledge, knows what the technology can and can't do. [...] Which is another factor in the successful links we've had with [*enthusiastic founder*], where ... he's not overselling the technology. (Interview 1–2)

Nevertheless, there were considerable cultural frictions – i.e. relational distance – to overcome.

> And one of those principles is, it's kind of alien to a scientist, and that is to start with the product that you want to be selling and always work backwards with what you have to do. Most scientists are interested in producing scientific data [...]. And that's all very interesting and useful. But you don't make money with good ideas and you don't make money with good science. So, what I kept challenging our team to develop was ... if you have a product in the end in mind, you're going into a very competitive market. So, question number one is: Is it going to be cost effective? How much is it going to cost to manufacture? (Interview 1–2)

From the start-up team's perspective, the challenge was considerable.

> If you asked me, I would test this and that in the lab. Make it better, step by step. And suddenly, they gave us this work [...] We had to do a filter test. Filtrate a thousand times and check how much was left. [...] You get a catalogue of questions from the authorities. The authorities want to have this and that checked. Scientists would say, 'This is bollocks, this is nonsense, idiot work'. But it has to be done. (Interview 1–1)

Although the start-up company remained formally independent, the team was accommodated at the pharmaceutical company's plant in Paris as if it were an internal project team. They were equipped with laboratory space, which necessitated a degree of spatial separation from – competing – internal project teams. A certain level of conflict needed to be managed.

> And we also have a few people that were pretty clear would never be able to work with [*enthusiastic founder*], because they couldn't work with any other external company. They would always think negative towards anything going on outside. [...] They're very clever people, and we keep them to generate ideas, we don't have them involved in implementing them. And we deliberately kept them at a reasonable distance from [*enthusiastic founder*]'s team, because it would have been very easy for them to be discouraging of the work that [*enthusiastic founder*]'s team were doing. And yes, you have to manage these things. We had to move some people around from the laboratory to create the space for [*enthusiastic founder*]'s team. That impacted on four people. Two of them were quite happy, one of them initially was concerned that it impacted her project, and the last one even to this day thinks we were crazy doing this. (Interview 1–2)

The team was given several drugs to experimentally encapsulate with their technology, which then were brought into a pre-clinical development stage using the company's facilities. This high degree of time-spatial and material integration was the first opportunity for the founding team to experience an industrial mode of working.

> In fact it was the possibility for us to test our technology in-vivo within the framework of this project. That was the main point. Of course we also got a glimpse into what it looks like when we move from the lab to the GMP area, when we really need to create a clinic sample, perform an upscale [...], work under aseptic conditions. At this time, I entered cleanrooms for the first time. This was simply a very, very impressive facility to see. (Interview 6)

Although from a technical perspective the success rate of this collaboration was very high, none of the encapsulated drugs were brought into clinical development or the market. The reasons given for this were considerations of business strategy, specifically comparative cost-benefit advantages of different product development options. However, the technical validity, which was demonstrated during the relation, allowed the founding team to enrol new investors afterwards.

The team relocated to Berlin, where the enthusiastic founder's founding career had begun. Several national and international localities had been under consideration as future locations before Berlin was selected. Respondents provided several reasons for this choice. At this time – in contrast to the late

1990s when the initial technology-oriented idea was conceived – Berlin provided laboratory space at low costs as well as an institutional support structure. Furthermore, the enthusiastic founder still had personal acquaintances in the city because of his earlier employment at a Berlin pharmaceutical company.

Using the results of the previous collaboration as proof of feasibility, they managed to enrol a local public VC fund as lead investor to fund an independent delivery service platform. In Berlin's public research organizations, research groups were working with a focus on pharmaceutical product delivery and could be approached as experts by the fund's management. The investment manager in charge also had personally worked with pharmaceutical delivery companies and was familiar with technology specific evaluative metrics. Crucially, the previous British–French strategic partner company testified to the validity of the technologic approach pursued. It is noteworthy that, although company representatives interviewed repeatedly stressed the technology's general openness and potentiality for a wide range of applications, throughout the unfolding of its relational web, the technology's attachment to a pharmaceutical application was reinforced in a cumulative way.

The public VC fund assisted the founding team in setting up shop and engaged in IP-related negotiations. It also managed to create an investment consortium with a German VC fund with an interest in life sciences, but, given its relatively small size, not in product development. The consortium funded facilities, staff and future operations. While the technology's validity in general could be demonstrated, the question of management strategy arose once again on the side of the investors. The team still operated under a very science-driven logic and was more motivated by enthusiasm for the technology's general capability and less by a specific market case. The team itself had not changed since the original company founding.

> Mr [*name of enthusiastic founder*] is a very extroverted person, who is good at selling what he does. [...] He is a good networker, but he is not the one who will turn unwritten agreements into a contractual framework, and who is capable of negotiating a contract with another party and close it. He is the one who initiates the first contact, rouses interest and inspires people. This is his distinguishing characteristic, apart from the scientific quality, which is undisputable. So, we said we need somebody who is capable of turning the customer contacts into jobs. (Interview 1–4)

Several attempts were made to install the formal position of a 'business developer' in the company and to fill it with persons recruited from either the investors' or the team's inter-personal networks. The aim was to instil a more market-centric, strategic perspective in the company. According to the business development strategy pursued under the fund's strategic supervision, the company was to establish itself as a service provider ('platform company') to pharmaceutical companies – to develop formulations either for new drugs or as improvements on existent products going generic ('product life cycle management'). Own product developments, as they were considered by the enthusiastic founder (e.g. by licensing a generic product and bringing it into clinical development with a new formulation), were outside the investment's scope and scale. However, none of the selected individuals managed to build a working relationship with the founding team. Finally, one of the three team members was chosen to fill the role. He was complemented with an experienced external 'sparring partner' with whom he would collaborate and thus learn from him.

Since then two joint investment rounds took place. Over the following years the company remained in its original state as a 'technology demonstrator'. The technology's basic ingenuity and validity were confirmed multiple times – by pharmaceutical companies who issued small-scale test runs for small fees and, on one occasion, through a market specific innovation trophy. However, no pioneering customer was willing to invest a relevant sum in an actual product development deal. More recently the company invested in upgrading its production capacity: By establishing GMP production standards it became capable of directly supplying the volumes and qualities needed for preclinical and stage-1 clinical trials, effectively reducing the relational distance to the practices of pharmaceutical product development. The application specificity for the field of pharmaceutical delivery was thus reinforced once again. However, till the end of data collection no deal with a pharmaceutical company was finalized.

In case 2, SYNTHESIS, the technological basis for a platform-oriented start-up company in Berlin was a procedure for synthesizing very large numbers of different peptide molecules and systematically testing their interactions with other molecules. This procedure was materialized in a synthesis robot, several of which were bought by the company founder (who was not the inventor, but, as a leading peptide chemist, a sophisticated user). The technique had been established in the epistemic community of peptide chemistry as a tool for developing and producing tailored substances for research in fields such as immunology. In the company, the synthesis machines were incrementally improved. While the basic architecture remained the same, individual components (such as steering software) were substantially increased in performance, increasing overall efficiency along a set path.

An R&D scientist at a Berlin pharmaceutical company and a personal acquaintance of the start-up founder acted as a pioneering customer: The pharmaceutical company used the technology in one of its development projects. Thus, the performance of the newly established company's molecule synthesis platform was evaluated. Both partners shared a background in peptide chemistry and believed in the value of tailored peptide molecules for pharmaceutical development. Although the project was not pursued further in the customer company, a proof of feasibility was established, adding credibility and reputation to the platform company.

The scientific community remained the main customer group, but pharmaceutical companies and their research departments also exerted demand. The commodity offered by the platform company consisted of a) the design of peptide molecules with particular properties (imitating organic molecules, but also variegating them) and b) the synthesis and delivery of small amounts of the respective substance for research purposes. Here, the production of molecules and the experimental process of locking them in identities and roles were highly intertwined. The product on offer can be described as a discovery service: the establishment of molecular properties and interactions as a potential starting point for pharmaceutical product development.

After acquiring IP of the Braunschweig-based technology and setting up a company in Berlin, he adopted a 'helicopter approach' to leading it. This included hiring a professional manager to handle day-to-day operations and maintaining a position at Charité University Hospital. For several years the company remained in its state as a low-risk, service-oriented technology platform in open exchange with the epistemic community of peptide chemistry. Only when the founder decided to use the company as a base for a more ambitious product-development project, did he take charge of management and initiate a transformation. Instead of developing a new drug on the basis of the existing technology platform, he licensed an unused substance from a German pharmaceutical company. Being peptide-based, the substance still matched the in-house technology. Having collaborated with several pharmaceutical corporations during his academic career phase in the United States as well as in Germany, the funder was well connected in the industry. He hired a former executive from the company, from which the product development had been licensed, to lead the development project.

In the first years of its existence, the company had not required VC since it had produced cash flow from the outset. With the envisioned pharmaceutical product development, this changed. Thus, the history of VC engagement in this case began in 2001. Embedding the product development in a hybrid business model while at the same time keeping both 'assets' separate regarding accounting, management, team affiliation and routine practices, was a strategy to attract venture capitalists in the aftermath of the financial crisis of 2000 to 2001: Investors now sought to evaluate assets separately. In 2001, the first financing round was realized by the Berlin state investment bank. The next investor was a German VC fund. Over the curse of several years, the investor consortium grew in size and spatial reach, from local to national to international. Until the company went public in 2005, during a peak of renewed stock market optimism shortly before the next financial crisis, a total of four financing rounds were realized, involving both European and American VC funds. As in case 4 BIOMARKER, investors did not exit at the time of the IPO but remained invested as stock holders. The main focus of these investments was the pharmaceutical product development and not the platform, which only changed incrementally. The drug in question could potentially be used for various indications. However, as other cases have shown, a

VC-funded product company cannot pursue several product developments at the same time. As in case 5 AUTOIMMUNE, an application falling under the orphan drug framework was chosen, because this framework provided an easier market entry path, but also higher margins.

> Indeed in the beginning we, too, wanted to develop this as a drug for cirrhosis of the liver, because the investors said, 'This is shit, such an orphan drug, the market is much too small'. All of this changed completely. Big companies go into the orphan drug market today, too. Because you can sell these drugs insanely expensive. (Interview 2–1)

The company's IPO had pronounced consequences for the technology platform. Firstly, for the longest time, the platform had been part of a larger, relatively open ecology, which crossed company boundaries. In the community of peptide chemists, similar synthesis machines had been used and exchange had taken place. The original manufacturer, a Cologne-based engineering company, had developed new variants of it. Up until the IPO, a comparatively liberal and informal regime of intellectual property (IP) ownership had been in place. With the IPO, the external boundaries of the company had to be drawn more clearly, and more strict delineations of IP had to be put in place. Additionally, being stock market listed, the company's investor relations were no longer based on negotiations in an exclusive circle, but subject to more formal rules, such as the laws against insider trading.

> [*The investors*] want to know things they are not allowed to know. They would have loved to have all the results or intermediate results of the clinical trials. You are not allowed to … This is a thing when you bring a company public; all of a sudden, communication within the company is completely different. This is difficult to understand for many people because they are not allowed to tell employees anything anymore. It's not possible. Otherwise, one of them starts buying or selling stocks because they have insider information. And the CEO ends up in prison. […] [*You need to go to*] conferences, show yourself. [*The investors*] want to look you in the eye. Like, 'Well, how's it going?' That's when it starts. You are not supposed to tell them anything. You cannot favour someone at the stock exchange. But of course they try. They ask a thousand questions and watch whether you twitch or change your posture. And then they try to read something into that. (Interview 2–1)

The shareholder interest was now the dominant consideration in company strategy. Combined with the volatility of capital markets, this meant a substantial narrowing of scope for strategic decisions. During the admission procedures for the pharmaceutical product, the company suffered substantial losses in stock value. At the time (2009), the CEO received an offer from a British pharmaceutical corporation, which specialized in orphan drugs, to buy the company at a highly attractive price. In this situation the sale was inevitable because every other course of action would have been a grave neglect of shareholder interests. In a stock market listed company this, like insider trading, can become a matter of criminal law.

The British pharmaceutical corporation was only interested in the drug development component of the company. As a consequence, the non-human representations of knowledge (patents, technological procedures, documents) were integrated into the corporation's business development structure, but the jobs associated with the drug development at the biotech company itself were lost. The technology platform was sold separately – a transformation made easier by the previous separation of assets.

It was bought by a relatively young German biopharmaceutical company, which attempted to become an independent, FIPCo, and therefore required a dedicated molecule synthesis site. This new company was organized as a holding, which was financed by SOUTHWEST FAMILY VC (see Chapter 2). Hence, the new owner had qualities of a strategic investor. The platform remained in the same locality (Berlin) but moved to a dedicated technology park. The option to relocate the platform to the buying company's location (Mainz) did not occur to participants. In this way, the technology platform was, as it were, re-embedded and at the same time became an element in SOUTHWEST FAMILY VC's project of building a new, fully integrated German biopharmaceutical corporation from scratch.

In case 6 STEM CELL, an initial investment consortium surrounding a technology driven platform company had been created in the period from 2002 to 2006. The consortium consisted of private investors, the original founders, KfW and the local *Sparkasse*'s VC office. An executive, who later entered the company, described both this consortium and the founders as intrinsically motivated, but inept regarding the necessary transformation from a scientific, experimental assemblage to a commercial business model. In this early period, a Land (state) VC fund had been approached, but declined to invest for that reason. However, the fund had knowledge of the company and its technological approach, which it considered promising. As in other cases, investors initiated the search for a new leadership. A new CEO was found: an executive with a history of both company-founding and work in pharmaceutical corporations. Under his leadership the company moved away from the original science-driven, open experimental approach and towards a more standardized and market-oriented mode of working. While the change was profound, it was described in less confrontative terms than, for instance, in case 4 BIOMARKER.

> The most interesting thing which I liked in this company when I joined was the scientific solidity. These are absolute experts, they are very frugal, they work very efficiently, fast, with minimal means. And what fascinated me: in the four years I have been here, no major misses of timelines have occurred. (Interview 6–1)

After the new CEO had initiated the 'market turn' in 2008, the public VC fund was ready to invest, along with a mid-sized German VC fund with a specialization on life science and IT, but not pharmaceutical product development. The two shared the lead. As in case 1, ENCAPSULATION, the company attempted to find strategic partners who would aid in advancing the platform's capacity and invest in a product development project (a human protein-based drug created using the company's stem cell cultures). Despite initial interest, no pharmaceutical corporation opted for this engagement. Over the following years, two additional financing rounds were realized, involving the two new investors, an additional private German VC fund with a similar specialization, as well as KfW and the original private investors. The latter participated in each round, despite substantial capital dilution. KfW remained the largest single investor and thus added weight to the unfolding investor network (and inertia to the innovation path). In total, above €20 million were invested over a period of six years.

In this constellation the company developed into a highly specialized service provider for stem cell-based production of genuinely human organic molecules for pharmaceutical application. Several national and European partnerships with 'build-up partners' (bringing knowledge in stem cell cultivation and GMP-certified production capabilities) were established to create this capacity. As in case 1 ENCAPSULATION, own product developments could not be realized within the consortium. The company pursued at least one early stage product development project, a vaccine, which from a scientific and technological side appeared as a natural and fitting application of the technology and thus was assigned worth by the scientists involved. While this project was perceived as a natural element of the company's overall profile *from the inside*, the prevailing outside view (by leading investors) was that this was not the case. The boundaries of the innovation were thus subject to negotiations and tensions between investor and investee side. In 2013 a second company was spun out of the original one to pursue the project separately. It received seed funding on a small scale (less than €1 million) by one of the private VC funds, while the mother company's delineation as a platform company was reinforced.

In contrast to case 2, SYNTHESIS, the marketable service in this case was positioned further higher on the 'food chain' of pharmaceutical product development. Its main focus was to make the production of *known*, genuinely human organic molecules feasible so they could be used in therapy. The customers were pharmaceutical companies who wished to establish a particular molecule as a pharmaceutical product. The service consisted of identifying a task-specific cell line out of the master cell bank, to cultivate it with specific media, to make the cells produce the compound (using viral vectors) and to establish the capability to do so at the customer's site. The customers would also buy a license to use the technology. Hence, knowledge was commodified by exploring, in practice, how cells can be made to behave in a particular way, by mobilizing the knowledge and by selling IP.

An entirely different form of platform was encountered in case 3, GENE FUNCTION. Here, the technological basis for company funding was the ability to test the actual functions of genes by observing the metabolisms of plant cells (using gas chromatographs as detection devices). Hence, organic materials were fed into the procedure as samples, but the only output was information (numbers on screens and documents). As Sunder Rajan (2006) stresses, information can be commodified through IPRs. This was the primary purpose of the platform which was created under the guiding influence (as well as money and property) of an investing German chemicals company. As described in Chapter 3, an old industrial building was emptied and infused with all the technical structures needed to support the analysis procedure. From the outset the assemblage was intended to function at an unprecedented performance level.

While other enabling technologies like robotics and gas chromatography were required to be state of the art and were provided by or via the industrial partner, the key challenge was computing. Sunder Rajan (2006) highlights the important role of databases as non-human actors and intermediaries in biotechnological knowledge creation. In this case, databases were an essential component and a bottleneck. In order to master the magnitude of data, radically new solutions had to be found.

> [...] Key technologies we needed back then were not existent. They simply did not exist. [...] In the first years, we worked with the central development department of [large computing company] in Palo Alto, California. The boss came here and said, 'What do you need?' 'I need this, this and this'. 'Easy!' Nine months later they came back and said, 'What you want cannot be done'. [...] Okay, try to process the 150 million data files we have per year. There is no technical solution, none at all. It did not exist. Even today, it only exists very rarely. I mean the technical solution to manage something like that and integrate the data from these different data files and to continue processing them as knowledge. There was no such solution. There also were no databases which could deliver something like that. [...] We then did it with another very large IT company, [name], based on our clear structure. (Interview 3–1)

Another IT-related challenge was the need to create software, which was able to recognize patterns, again processing huge amounts of data.

> And with [name] institute, which is roughly the pendant to German Fraunhofer Society, a high-tech, semi-public research institute in the USA which does space research, and with several others [...] With these two partners and our little shop here, we [adapted] a piece of software which [name] had created for the CIA, which the CIA then had released in 1996 or 1997, with which the CIA in 1997 was able to scan 50 million phone calls for speech fragments. That was 13 years ago. We took this

software and developed it further for biotechnological applications. Because we are talking about genetic patterns, we are talking about metabolic patterns and things like that. And we are confronted with such a flood of data, in which you have to get sense, where you have to recognize patterns. (Interview 3–1)

Case 3, GENE FUNCTION, is special in the sense that an experienced industrial manager, capable of initiating an industrial mode of working, 'came' directly with the first – and only – investment. In advance of the engagement, he had worked in an R&D plant of the investing German chemical corporation in the United States. He was sent to lead the newly founded joint venture of Max Planck scientists and the aforementioned company. To this purpose, he was granted access to the company's organizational and technical resources.

And [*chemical corporation*] created this framework around us. They said, 'If we can deliver anything to you, we have competences. [*name of executive*], go, pick and choose'. And that is what we did. (Interview 3–1)

Hence, the enthusiastic founding team, who went into the new joint venture company, was integrated into the new market-oriented logic from the start instead of being replaced at a later stage.

Mr. [*executive*] professionalized us right from the start. [...] We made plans, we used professional planning tools, we had regular meetings with agendas and minutes. [...] So, we immediately adopted, I think, this industrial character of work. (Interview 3–3)

At the same time, the chemical company had a genuine interest in building a radically new technology platform, based on the desire to catch up in the green biotechnology sector, after lagging behind American competition for roughly a decade. Therefore, scientific enthusiasm and technology-driven thinking were appreciated in the situation. An unusual degree of improvisation and open debate took place, which was rather reminiscent of a classic start-up company.

Of course these are different cultures. Industrial culture is different from academic culture. And I believe that what you can see in this example here is that, if you have a common goal and develop mutual trust, suddenly out of these two cultures a mixed culture emerges which is highly productive. [...] This is technology and intellect. [...] Here in this room we sat, sometimes through the night till five or six in the morning, dead tired. And thought up five or six solutions which we tried then. (Interview 1–2)

The overall framework was, however, unmistakably industrial.

And then, yes, this was important: we positioned ourselves with four technical centres from the start. The management team consisted of these technical centre leaders and we also immediately established the matrix organization principle with project leaders and technical centre leaders. So, whenever we hired a new person, the game began anew to say, 'This is how we are organized, this job is in this technical centre, it does this and receives input from the previous technical centre and hands it down to the next'. The people experienced how we think, how we organize, during the interview process. I think in this manner, we also selected people that would fit in here and would not try to continue their academic career. (Interview 3–3)

The platform's original purpose was to perform a predefined set of analysis operations, specifically to 'scan' a variety of organisms (plants) for relationships between cell metabolism and genome. The results were patented as a basis for product development.

Build-up, this is where many companies fail. We went till the end of this project and were extremely successful, otherwise we wouldn't have this [*American agro-biotechnology company*] partnership. We patented about 150,000 genes. We submitted huge patents. (Interview 3–1)

This took roughly five years. Upon reaching this milestone, the decision was made not to disassemble the platform, but to both diversify and intensify its utilization. A second subsidiary company was founded to exploit the platform's analysis capacity on the pharmaceuticals market. The platform's technologic assemblage was a materialization of a new paradigm in functional genetics research, relating to both plant and animal (hence human) cells. Therefore, it could be used to perform analyses of the effects of various substances on different types of human cells.

Pharmaceutical companies could use this type of analysis to test drug candidates before they went into pre-clinical development. An additional interface was added and specialists for the pharmaceutical market were hired. Crucially, the platform as a physical entity and non-human actor stayed the same. Instead of plant seeds, human blood or urine could be entered into the system in separate laboratory spaces and the results were to be assessed by medical experts. But the platform behind the interfaces performed both operations, and at the time was unrivalled in performance in both fields. Hence, a material association of the pharmaceutical and the crop breeding market as well as an influential position in both was created.

Additionally, in the mid-2000s, the chemicals company entered into a strategic partnership with a leading American agrochemical and agricultural biotechnology corporation. In this partnership, the platform's purpose remained producing IP in functional genomics. Hence, it was to become a key asset in joint efforts of product development, i.e. in developing new genetically modified crops. Between the German and the American corporation, sales turnovers emerging from jointly developed products were to be shared according to a predefined split. Both of these two trajectories of knowledge exploitation share the characteristics of a large-scale strategic or industrial investment logic: the ability to materially build a high-performance technologic assembly based on a smaller-scale experimental assembly is combined with a strategic and shareholder-oriented approach to creating influential market positions. Through the creation of a very powerful actor-network (the platform) described above, two markets – pharmaceutical discovery services and the creation of new crops – were connected and a domineering position was assumed in both.

4.4 Transforming innovation networks

The cases referenced in the last sections show that biotechnology innovations undergo deep transformations on their path from an initial materialization (typically involving a start-up founding) to a viable business case and finally the market. They change in terms of people involved, required skills and dominant type of knowledge work, formalization, division of labour between sites, organization of knowledge work across physical distance and guiding rationales. This transformation often presents itself as a severe cultural clash. Whether and how a novel idea is put into practice on a market rests unforeseeable, sometimes accidental and very rare opportunities. These events, however, may not be entirely random in nature. They depend on conditions which can be considered structural forces: the market power of large corporations, regulatory requirements, cultures of usage to name a few. Navigating this line, discovering and co-creating the market opportunities which emerge out of the slow movements of 'big leviathans' (Callon and Latour 1981) is yet another typical form of agency and relational work. I will refer to it as 'pipeline building'. It is often more distributed across a number of individuals. Nevertheless, in many cases individuals appear who are credited with leading the transformation and cultural change required for a successful innovation. They will be referred to as 'pipeline builders' or, as a more technical term 'anticipating and transformative entrepreneurs'.

The metaphorical denomination of 'pipeline building' carries two key associations. In the 'buzz and pipelines' literature (Bathelt, Malmberg and Maskell 2004), 'pipelines' represent trans-local relations of knowledge commercialization, in which knowledge is mobilized from the localized 'buzz' which spawns new ideas. Pipeline builders work to achieve the transformation from localized to globalized knowledge. In this sense, Birch's (2012) two key dynamics in biotechnological knowledge creation are visible in every single innovation process. In a more general sense, a pipeline has a starting point, an end point and a route which it follows between the two. When a pipeline is built, all three are chosen in a very careful, highly selective and highly contested process. Therefore, the metaphor brings to the fore the selective, possibly path inducing nature of this type of work. Some elements are connected in new, far reaching ways, whereas others are deliberately left out. The selective dynamic of making some connections very strong and abandoning others is characteristic for the transformations which innovation processes must undergo. Identifying the most productive route, anticipating potential obstacles, pitfalls and contestations, enrolling a far-reaching network of stakeholders, being conscious of strategic implications and placing feasibility above everything else – all these virtues are facets of the pipeline builder's professionalism. Quite literally, the individuals involved are concerned with developing product pipelines in biotechnology companies.

A wide range of actors come into consideration as 'pipeline builders': company founders, investors, industrial executives, appointed CEOs. While it is less easy than in the case of, for example, the establishment of an enthusiasm-led company to identify the concrete individual, the agency is often more distributed, key individuals appeared in most cases, such as the experienced CEO equipped with industrial, academic and entrepreneurial track records in case 5 AUTOIMMUNE or the private investor in case 7 CANCER IMMUNE, likewise with academic, industrial and entrepreneurial merits (Table 4.1 gives an overview of pipeline builders across the cases). Crucially, pipeline building is not a clear-cut, institutionally or organizationally defined role which could be filled by an individual trained in one logic alone. The individuals observed had unique mixes of backgrounds and skill sets. They were rare and difficult to find.

A variety of concepts each shed light on aspects of pipeline builders' contribution: They can be understood as brokers (Obstfeld 2005) and 'boundary

Table 4.1 'Pipeline builders' across the cases

Case	'Pipeline builder'
1 Encapsulation	Pharmaceutical executive
2 Synthesis	Former university hospital working group leader and company founder (he hired an industrial executive to implement his strategy; also the case's 'enthusiastic founder')
3 Gene Function	Chemicals company executive
4 Biomarker	Diagnostics company executive, among others
5 Autoimmune	Biotech and former pharmaceutical executive, former founder (in an earlier phase also participated in 'place making')
6 Stem Cell	Biotech and former pharmaceutical executive, former founder
7 Cancer Immune	Private investor, founder and consultant, former scientist (also the case's 'place maker', see Chapter 5)
8 Neuron	University hospital scientist working in a collaborative relationship with a pharmaceutical company (also the case's 'enthusiastic founder' and 'place maker', see Chapter 5)

Source: Own design

spanners' (Tushman 1977) as they build linkages and translate between culturally distant groups and organizations. Some act as 'intrapreneurs' (Gap and Fisher 2007), initiating and managing change within organizations while opening them up to external influence (e.g. the pharmaceutical executive in case 1 ENCAPSULATION who accommodated a start-up team in his company, fighting internal resistance, and at the same time instilled an industrial style of working in the team). Given the repeated occurrence of mixed biographies combining excellence in two or three fields, pipeline builders can also be characterized as multiple insiders (Vedres and Stark 2010). However, *acting* as a multiple insider appears to be less relevant than *having been* in different contexts throughout one's career.

'Pipeline builders' can be described as professional managers who cultivate a sense of openness and cognitive independence. Their logic of valuation combines an intrinsic appreciation of classical, individual entrepreneurship and pioneering technology with a high professional regard for the more mundane necessities of product development: standardized testing, documentation, SOP, admission procedures and production, the integration of user preferences (however banal) and the organizational mechanisms needed to connect all involved practice contexts. In this idea of worth, the unbridled enthusiasm for a particular technology often displayed by founders is replaced by a much soberer sense of an idea's feasibility beyond creative environments. It represents a more cautious and strategic approach to biotechnology innovations.

The aspect of anticipation refers to the needs and preferences of gatekeepers like regulators, corporate strategists or investors, which need to be carefully assessed in advance if a new technology is to be made into a marketable product. Establishing a product on the market is the goal. Activities which facilitate this goal, like standardized testing and investor enrolment, are of instrumental value. This instrumental value is associated with an intrinsic, quasi-moral valuation and appreciation of managerial professionalism, in which the ability to organize a process takes precedence over scientific genius. User needs and a pragmatic, usable and marketable solution take precedence over a scientifically perfect one. The individuals involved understand the need to distribute the activities of product development across various formalized roles and actors and thus have a less personal attachment to technology-centric ideas.

However, they also have an appreciation for start-up entrepreneurialism, have an affinity to risk and cherish individual autonomy. This is evident by the fact that throughout the cases, strategists, in addition to careers in or close to industry, also have histories of company founding. Like place makers, they typically founded start-up companies from an occupation in academia or industry earlier in their careers. Some founded several. Following an idealistic and sometimes personal mission is one of the motivations to do so. One entrepreneur explained:

> Then of course I saw it in my family and elsewhere, time and time again, how terrible it is when people more or less lose their personality through this disease [*Alzheimer's*]. And that is of course a motivation to do something useful in this direction, and not to keep researching in a more esoteric fashion, as I did earlier. These are, as it were, personal reasons. (Interview 6–1)

A different respondent described the loss of a child to an untreatable disease as a key motivation to accept responsibility for a biotechnology company which developed a therapy for that disease. Mirroring personal passion for individual projects, respondents often reported dissatisfaction with overly bureaucratic working environments in large industrial corporations as a motivation to 'switch sides'. They felt too distant from actual work 'on the ground', i.e. from developing ideas and advancing projects.

> When a colleague [in the pharmaceutical industry] told me one day, 'I know I made it because [...] nowadays I draw up more organization charts than scientific formulae', I knew I had been in there too long. (Interview e–2)

Across the cases, respondents agreed that being socialized in a corporate context *alone* does not suffice to lead the necessary transformations in an innovation process:

> If you imagine that somebody has worked in the pharmaceutical industry for 20 years [...] then they usually have not contributed to innovation, but rather applied the classic pharmaceutical model which has existed for 50 years. [...] Generally, people are not used to taking risks. Generally, people are not used to fighting for a risky, innovative new project. In the pharmaceutical industry, they are not rewarded for taking risks, but for not making mistakes. They do not make careers out of continuously touching risky projects. They make careers out of not making big mistakes. (Interview 5–3)

Transnational mobility and transnational work feature frequently as important elements in transformative entrepreneurs' backgrounds and careers. While not all 'pipeline builders' had a transnational career, organizing processes of knowledge creation which cross national borders was part of the tasks of every single one. Territories and their institutional or cultural properties were frequently subject to organized reflexivity – their suitability as markets or research spaces, their specific challenges and potentials. Furthermore, knowing how to use the materiality and physicality of space to organize collaborative knowledge work over long distance, to experience difference and to manage or exploit it, was an integral part of their practical competence (see Helbrecht 2011). Such practices include organizing project teams across multiple sites and territories using electronic media and temporary co-presence, standardized documentation of work and technical virtualization of working environments.

Typically, 'pipeline builders' had a prior connection to the networks of acquaintance and reputation surrounding innovation processes, especially the investor networks. The most frequent case was that early investors perceived the need for a strategic reorientation of the innovation project they had engaged with and began searching for more suitable leadership personnel. 'Pipeline builders' were then hired as new CEOs or as business developers, often to replace enthusiastic founders in this function. They typically experienced this offering of a new mission as both a personal chance, which allowed them to reconcile both of their 'souls', the professional-yet-dull, industrial one and the more adventurous, start-up-oriented one, and to achieve something good for society in the process. In terms of payment, this was usually a losing deal, even more so if the increased risk of failure of a start-up vis-à-vis a large corporation is taken into consideration. Frequently 'pipeline builders' (if they were hired as CEOs) were required to personally co-invest in the respective company, thus 'incentivizing' them to seek commercial success and minimize risks.

In almost all cases, the integration of a 'pipeline builder' into the innovation process required a change of location: either the actors would move to the locality where the pre-existing socio-technical assemblage was located – the usual case – or the innovation itself (people, equipment, data – case 1, ENCAPSULATION) would relocate. These relocations were either inter-regional or even international. This circumstance reflects the high specificity of the respective match: 'pipeline builders' are ready to move over long distances to take on a very particular challenge. Like enthusiastic founders, they are readier to do so if the move can be aligned with their overall life-cycle-based mobility and the associated search for personal value. However, since the individuals in question here tend to be more advanced in their careers, other motives (such as family rather than excitement) enter the equation. In one case (case 5, AUTOIMMUNE) the 'pipeline builder' returned to his home region in southern Germany for a last assignment before retirement after having spent several years leading a Berlin start-up company.

From the other side, too, these matches are highly specific. Early investors and other stakeholders employ existing inter-personal networks of reputation and personal acquaintance to find a candidate who is a) suitable (representing the appropriate mixture of field specific experience, risk affinity, business acumen and personal human qualities as a manager), and b) at a career stage which indicates availability. The driving force behind the search for a new leadership is the commercial interest in transforming a more experimental enterprise into a viable business or innovation project as well as the inability of any actor already present in the enterprise to effect this change (including investors).

In most cases (all except one), the 'pipeline builders' who were successful in transforming the innovation had some form of previous attachment to the respective company as well as a personal appreciation for or intrinsic interest in the respective project pursued. Business developers or consultants, however formally qualified, typically are not able to fill this role. Hence, classic venture capitalists, as long as they don't engage with innovations in a very early stage and in an 'intimate' way, are severely handicapped in this regard.

It is productive to think about the activities of 'pipeline builders' in the same terms as the activities in earlier stages have been described, i.e. in terms of translations, of struggles to replace existing relational ties and shared purposes by new ones. This means scrutinizing the problematizations (suggested orders of roles, identities and relations under a shared purpose) they propose, the ones they reject and strive to alter, as well as the devices of interessement they use and the ensuing struggles and trials of strength until ultimately enrolment is achieved.

The key point of attack is typically the central roles of enthusiasm-driven founders in the translations they have proposed. In such a translation, an entire innovation is defined by a new techno-scientific concept whose value lies in its novelty as an idea and in its open, yet unspecific *promise* of applicability. Enthusiastic founders claim an exclusive competence to define this core technology. In time, space and matter, this claim is expressed in the establishment of a localized network of artefacts, technical procedures, practices and interactions of human and non-human actors. This nucleus is stable, yet dynamic. It is first and foremost unique, as it includes open and experimental procedures, custom-made artefacts and humans with an individual, personal history both with each other and with the artefacts and procedures. Hence, it evolves incrementally and idiosyncratically out of ongoing, localized interactions. Patenting, founding a company, enrolling investors – all these activities are directed at stabilizing and legitimizing this network as an independent entity – outside the environment in which it was conceived, such as a public research project or a service arrangement.

With the establishment of a 'pipeline builder' in a leading position – sometimes replacing the original founder, – the first 'strike' in the 'attack' on the early, science-driven translation is already carried out. However, the groundwork, dealing with the further implications, is left to 'pipeline builders'. It is up to them to propose a detailed new definition and delineation of the innovation, an ascription of value and purpose, and a set of roles and relations that come with it. In the new alignment, the key value of *concrete materializations* of a technology is their suitability as commodities in markets. This means that actors need to be identified (and later enrolled), who can pave the path to a market entry: pioneering customers, who are ready to adopt a new solution, regulators, opinion leaders in the user community and others. As each market and each sub-section of a market is different, this proposition of roles and relations needs to be rather specific. It also needs to include the more mundane aspects of product design.

However, as stated above, market entries are opportunity-driven. At the point in time (and space) when a 'pipeline builder' enters an innovation process, the concrete opportunity which will in fact allow a market entry is in most cases far from visible. Until, for example, regulators or potential users can be approached, several development steps need to be taken to ensure that the technology's key elements will work together outside the experimental lab context. In the case of pharmaceutical product development this means, for instance, pre-clinical and early clinical trials. This is to a certain degree a catch-22 situation. The specifications are needed as interessement devices – datasets, PowerPoint slides and brochures stating things like 'easy to use, proven effectivity' – to enrol market actors. It is, however, unclear whether these particular actors will be ready to adopt the new solution several years in the future, because the opportunity structure might be adverse by then.

To achieve a market entry, transformative entrepreneurs must reformulate the shared purpose of people working on an innovation project: from a science-driven exploration towards a formally structured product development project with clearly defined, formalized roles and a dramatically reduced level of scientific complexity and uncertainty (yet with an increased focus on regulatory uncertainties). The conflict appears obvious. However, the two logics usually coexist peacefully in separate sites of social practice. What makes the conflict relevant (or rather: existent) in innovation processes is the concrete situation in which it is experienced by participants: an identifiable group of individuals has initiated a project for which they claim cognitive, emotional and material ownership. The tight and exclusive relationship between people and a project is, or rather has been, the basis of progress so far. Now this ownership is challenged. The rationale for this challenge which is put forward is the further advancement of the very same project. Hence, initiators face the choice to either subordinate themselves to an identity-altering new order or withdraw.

One of the mechanisms to solve these conflicts was formal, hierarchical organization. Initially, the establishment of an organization, i.e. a company including a legal form with associated ownership rights and decision-making mechanisms, had often been a means to obtain independence for enthusiasm-driven founders. Company founding, along with patenting, business plans and other things had been a precondition for enrolling support. Despite this new institutional context, technology-driven founders had typically maintained their open experimental style, both with regard to scientific work and strategic decision making. Now, as a substantial transformation ensued, the formal organizational structure received the function of an interessement device to break up an existing order and to establish a new one. Frequently, the shift from the scientific mode to a product development mode in a start-up company was associated with a) a substantial reduction in personnel and b) a personnel exchange, during which more academia-oriented scientists were replaced by more business-oriented staff. Some founding members also took this dynamic as an opportunity to learn a new role and thus become 'biographic' boundary spanners. It occurred in almost all cases that members of the original founding team migrated back into academic positions as soon as they realized that product development work was incompatible with their professional identity. Furthermore, despite the importance of hierarchy and ownership rights in effecting the transformation, the social skills – empathy, respect for scientific professionalism and ability to convince – were also frequently cited as crucial 'devices' of interessement.

The personnel turnover, which is associated with a strategic turn in biotechnology innovation processes, sheds light on the way such processes are embedded in localities. Founding teams prefer to set up companies close to home. Furthermore, spatial proximity to previous sites of work (universities and other public research facilities) is an advantage as long as work on the innovation is science- and technology-driven, because exchange with former peers is simplified by physical vicinity. In addition, having the opportunity to go back to an academic occupation without the need to relocate is an advantage from the perspective of individual labour market resilience (Ibert and Schmidt 2014). From the perspective of a company in transition, a locality's attractiveness as a place to live and work is a relevant factor for the enrolment of new individuals who bring the specific qualifications needed for product development. Hence, while the original mode of location specificity of an innovation is gradually dissolved, the transition itself is highly location-sensitive.

The transformation of a central site in an innovation process (that is, central with respect to technology developments) is not an isolated incident in time and space. It is embedded in the unfolding of an increasingly sophisticated network of interaction, orchestrated across several specialized sites and localities. In this unfolding, the role of space is fundamentally different from its role during earlier phases. While idiosyncratic spatial settings, overlaps of practices and coincidental encounters have shaped early emergent socio-technical assemblages, space is now consciously functionalized. Sites and localities are selected and integrated based on their utility for a particular purpose. Frequently they are also transformed in the course of events. More distanced, specialized networks emerge. Practices of organizing knowledge work over physical distances become increasingly important. Product development is a spatially distributed, decentralized activity (see also Ibert 2007). Frequently new forms of centrality are brought into an innovation network in this way, represented for example by the sites of capital allocation (investment) and high-level corporate management.

The processes, during which such specialized networks are created, are not uniform. Substantial differences are visible depending on the business development strategy pursued in the respective case. Furthermore, there are movements in opposite directions. During early stages, enthusiastic founders often faced the necessity to separate themselves from inhibiting environments. Similarly, the creation of a network of actors across multiple places is a process of building ties and cutting them, of conscious linkage and separation.

Managing this time-spatial unfolding of relations is another challenge in pipeline building. It can be subdivided into two basic fields of activity: developing a business model and a pathway to the market including the building of appropriate trans-local relations and managing relations with investors to ensure funding.

Establishing a working business model is a mirror-imaging activity to the creation and maintaining of investment relationships. It is this two-way relational work which produces the pathways of commercialization in biotechnology innovation processes. Searching for new investors and maintaining the relationships with existing ones (the 'shareholding relationship' as it is denominated by Ibert, Müller and Stein 2014) is by far the most time-consuming activity of a biotechnology CEO, as one respondent explained. Throughout the cases, development processes turned out costlier and protracted than anticipated by both enthusiastic founders and 'pipeline builders'. The permanent need to convince already enrolled investors to extend their commitment, while at the same time struggling to find new investors, is a repeating motive throughout the cases. 'Cliff hangers', i.e. phases with no or almost no funds repeatedly preceded the realization of new financing rounds.

> Yes, it's difficult, terribly difficult. I heard a presentation by a guy from [company] [...] who said, 'You have to kiss a thousand frogs until one turns into a princess'. Yes, as absurd as that sounds, it leads to the truth. In fact, we were constantly almost broke. The money sufficed, and then we ran out of money and had to get new money. And then every time we made it by one or two days [before bankruptcy]. But this is very exhausting, yes. (Interview 2–1)

The sometimes desperate search for investors while being under pressure to pay employees, the 'standing with your back to the wall', is one of the defining experiences of the anticipating and transformative entrepreneur. Respondents frequently referred to this shared experience as a form of baptism of fire, an initiation into the practice of biotechnology entrepreneurialism, and used it to separate its 'true' practitioners from would-be entrepreneurs (such as consultants and investors with no own entrepreneurial experience).

In addition, the necessary steps towards the market cost money. The funds provided by early investors are typically sufficient for a general 'proof of concept' on a small scale. Therefore, additional investors need to be enrolled. The required funds for product development in biotechnology often exceed the original seed funding by a factor of 10 to 100. This circumstance can necessitate a second layer of commodification, or rather assetization (Birch 2016). 'Pipeline builders' create problematizations in which two 'obstacle problems' are connected: the investors' presumed desire to invest funds in a profitable way and users' and regulators' desire to have a new, better (more effective, easier to use, less troubled by side effects) product in the market. The resulting structure of roles and relations is more complex than the earlier, technology-centric one. Both users and investors assume more active positions rather than being mere allies or recipients of technological genius. While enthusiastic founders lose their central position, an even more central place emerges for the 'pipeline builders', who negotiate between very different worlds, building links across large relational distances.

What makes this position particularly challenging, but also in some way creative, is the need for ongoing reflexive adjustments between events on the capital-market side (venture capitalists seek exits due to limited fund run-times, different investment priorities clash in a consortium etc.) and on the product development side (unforeseen events during testing, a potential lead customer pulls out etc.). Frequently, the necessity arises to redefine the innovation and its boundaries in order to respond to situational circumstances, while at the same time, stability needs to be maintained to calm investors. Full enrolment is achieved, and hence, the translation is completed once a product is established in a market. Until then, the translation is in a liquid state, in which propositions of adjusted problematizations alternate with struggles of interessement and dissidence.

In very general terms, under the new emergent translation, an innovation is no longer structured and delineated according to the 'natural' boundaries of a new techno-scientific approach (for example, in case 1, ENCAPSULATION: all substance encapsulations which can be performed with a particular encapsulation technology). Instead, it is delineated by the boundaries of a coherent business model with a specific risk profile, specific return expectations and a specific pathway to the market. This means *in some cases* that material associations and interaction patterns are cut off. Cases 2, SYNTHESIS, and 6, STEM CELL, for example, have in common that they combine a technology platform[6] with one or several individual product developments (drugs). The both sides are technologically interdependent as the platform can be used to develop the product, and the scientists involved experience this exchange as stimulating and value-creating.

Under the influence of VC investors (and in case 2, SYNTHESIS, stock market regulations), 'pipeline builders' found it necessary to separate those two elements organizationally and install separate management routines and accounting frames. Investors needed to be able to assess risks and benefits in both branches separately and treated them as separate assets. This financial separation entailed the option to split both operations completely, that is, to create separate companies.

4.5 The importance of strategic investment

In addition to the finance relationship, other quasi-investment relations need to be managed. These will be termed 'strategic relations'. Strategic investors, usually corporations in the pharmaceutical or agro-chemical sector, take an interest in innovative technologies as a potential means to develop their own business. By investing early, they hope to gain exclusive access to a new solution at relatively low costs. This motivation strongly differs from that of capital market-oriented investors such as classic venture capitalists who seek an exit, i.e. a profitable disinvestment. One such can usually be realized after a sale on one of several possible markets: the market for mergers and acquisitions (M&A) in case an entire company is sold ('trade sale'), the market for licenses in case a particular product development is sold in a late stage, the stock market in case a company is brought public, or an end-user market where a company sells products and pays back its dues continuously. In an ideal-typical dichotomy, strategic investors seek a low price and a technology for themselves while venture capitalists seek a high price on a sale which is open to many bidders.

Strategic relations combine elements of financial investment with other forms of collaboration and value creation. Two relationships, which were described by Ibert, Müller and Stein (2014) fall under this category: the 'build-up partnership', in which technology suppliers help to build more capable, industrial technical assemblage, and the 'pioneer user relationship', in which potential customers provide resources to evaluate and test a novel solution under real life conditions. Both relationships were described as highly selective and the products of active, targeted search for the most fitting partner. Likewise, they were characterized as operating over long distances in physical space. In both cases the partners are usually industrial corporations. In practice, both relationships often overlap. Therefore, the term 'strategic investment relations' here refers to all relational arrangements, in which new materializations of knowledge and financial engagements are combined, and which contribute crucially to the preparation of a market entry. Such relations took various forms and occurred in various combinations and at different stages in innovation processes throughout the studied cases.

Strategic relations relate to mere financial investor relations in an ambiguous way: they create certainty, which is welcomed by venture capitalists. They do so by aiding in the creation of investment opportunities, by signalling proof and validity and by providing exit options. In this sense, they can be structuring elements, which enable VC investment (or disable it when they are absent as in the intermediate phase of case 8 NEURON). But they can also bring logics which contradict the order of venture capitalism into an innovation process. Finally, the relationship is not mono-directional, as VC investment or the stock market can also provide opportunities to create strategic relations.

In practice, the different logics, enrolment mechanisms and forms of material entanglement of financial and strategic investment are brought together by making use of very time sensitive opportunities. Furthermore, the divergent orders of worth are subject to constant negotiation. Venture capitalists in particular have developed varied strategies to deal with the divergence. Some adapt their action logics so as to interpret the formal and informal orders of venture capitalism liberally. New forms of venture capitalism emerge. Others assume a highly defensive position and drastically limit their engagement with biotechnology innovation processes. Venture capitalists also learn to appreciate the reduction of uncertainty brought into investments by strategic investors (industrial corporations), notably in the form of proofs of concept and feasibility they are able to deliver. In turn, strategic investors and public VC funds partially adopt the orders of venture capitalism as their own logics of value appreciation as a means to make themselves 'compatible' to VCs. Finally, 'pipeline builder'-type entrepreneurs use temporary constellations of investor participation and other temporal context conditions as opportunities to perform leaps towards new markets and financing sources, such as an IPO or a trade sale.

Understood functionally, strategic relations provide settings, in which a set of key transformations can be achieved: scaling and certification of both, analysis procedures and production processes, testing of feasibility under 'real-life' conditions, testing according to regulatory standards, executing other regulatory requirements and creating sales and distribution structures. In an ideal case, strategic relations enable a market entry. Crucially, they also contribute to innovations even if they are temporary or if they fail. In strategic relations, founders can get access to specialized sites and participate in practices which are otherwise closed to them. In some cases, strategic investments occurred very early, thus kickstarting the respective innovation processes. Only in one of them, case 3 GENE FUNCTION, the investment led to the intended commercialization. In cases 1 ENCAPSULATION and 5 AUTOIMMUNE, the innovations benefited vitally from very early strategic investments, while the investing companies did not participate in the innovations' commercial success.

In strategic relationships, innovations' emergent multi-site networks are connected to or integrated into the multi-site networks of industrial corporations. This form of integration is fundamentally different from a pure financial investment. In the latter, documents and people circulate. The 'home' sites of investors (often first- or second-tier financial centres) become new centres in the network and the logics which operate in them are implemented in the former key sites of the innovation. Investors are both allies and challengers of pre-existing translations and relationships.

In strategic investment relations, a richer exchange takes place, involving a greater diversity of human and non-human actors. Industrial corporations open themselves up and thus grant access to sites and, more importantly, practices which are usually closed to science-driven entrepreneurs. While biotechnology companies have the opportunity, at considerable expense, to hire CMOs or CROs, their practices and sites remain black boxes. Furthermore, a precondition of their enrolment is a predefined object of analysis or production, i.e. an envisioned product.

In strategic relations, this kind of certainty can be created through negotiation without it being present from the outset. The relations can take many legal forms: They can take the shape of an actual equity investment or that of a service contract or a strategic collaboration. In them, the partners usually do agree upon a pilot project and define terms, conditions and mutual obligations. In theory, there is a greater openness to respond to unforeseen events, to adjust and refocus the work, as it happened for instance in case 3 GENE FUNCTION. The reason is that industrial corporations are 'big leviathans', who consist of a large variety of actors, sites and practices. Some of these can participate in the more science-driven work on technologies and extend its reach or change its quality. Pharmaceutical companies, for example, have production facilities, animal testing sites, high performance arrays for gene sequencing, administrative capabilities for clinical trials etc. This bigger (compared to financial investors or contract partners) 'internal space' allows for a more flexible re-contextualization, both in terms of capacity and practical troubleshooting as in terms of evaluation. On the other hand, strategic relations can end in unpredictable ways when interests and majorities in the investing corporation shift. Cases 1 ENCAPSULATION and 4 BIOMARKER provide examples of such unforeseen break-ups.

In sum, enrolling an industrial corporation into an innovation process is productive to entrepreneurs in three ways: the financial component, the additional capability of powerful industrial actor-networks, and the reputation of having won a relevant ally. Depending on the circumstances, all of these contributions can remain valuable even if a project fails and a relation is broken up.

The questions as to who the actors of strategic investment are and how they can be enrolled are crucial. The category of 'industry' is a rather broad one. Throughout the cases, an interesting dichotomy could be observed. On the one hand, strategic investors are the 'usual suspects': international corporations in the pharmaceutical, chemical, agro-industrial (crop breeding, seeds) and diagnostic fields. These companies tend to be stock market-listed, very large (especially in terms of stock market valuation) and concentrated in the United States, the UK and a small number of other European countries. While Germany as a headquarters location is represented in the pharmaceutical field with only three relevant, yet comparatively small players (Bayer, Boehringer Ingelheim and Merck), it houses the world's largest chemicals corporation, BASF, along with many other leading chemicals firms. Companies such as these obtain very exclusive positions at the intersections of the capital market (via their stock market presence), the regulatory environment and end-user markets.

The enrolment of a strategic investor is an immensely ambitious undertaking and, if successful, an exceptional event. It requires the initiators, in most cases biotech entrepreneurs, to identify a company whose leading representatives perceive the proposal as fitting to the company's business development strategy *at this particular point in time*. Potential strategic investors receive proposals through a great variety of channels: face-to-face contacts at partnering conferences, scientific conferences and other occasions of temporary co-presence, via electronic and social media including companies' own websites and via existing relationships. In other cases, R&D personnel within the company approach leading executives bottom-up and initiate a search process for a new technologic approach. Company representatives stress that their spatial scope of search is global, and that spatial preferences or territoriality do not enter into the evaluative logics for external approaches. They also stress that there is no preferred or dominant way through which potential investments are identified.

However, the dynamics in the cases studied indicate that there are several potential constraints which make a successful union a more context-sensitive event. Firstly, strategic investors very often face the choice between an internal and an external solution to a problem. This choice bears the potential for conflict with internal product developers and proponents of 'homemade' solutions, as was evident in case 1 ENCAPSULATION. Managing this potential conflict is a part of business development practices which include external search. Scientists and technology-driven founders increasingly perceive large corporations, especially in the pharmaceutical sector, as marketing and sales organizations with few remaining internal research capabilities. From the perspective of complementarity, this would imply better chances for biotechnology companies to sell their solutions, as internal competition is reduced.

From a practical standpoint, the opposite is true. For biotech entrepreneurs, the natural communities of practice to engage with in an industrial corporation are those of R&D researchers. Several private and VC investors pointed out that they encouraged management teams of investee companies to seek personal contact to 'gatekeepers' (i.e. leading product development executives in the industry), and to establish face-to-face contacts with industrial R&D staff on science conferences. Preferably, these conversation partners share a disciplinary background and thus an enculturation in an epistemic community with the biotechnologists. It is up to the R&D staff in a company to evaluate an external proposition from a technologic and scientific perspective. If there is a shared background and possible shared experience with particular technologies, assessing and evaluating proposals is a more open, engaging and multifaceted act in which various uncertainties can be productively accommodated and differentiated valuations can be negotiated (compared to, for example, an assessment by financial investors).

On the other hand, if there is no shared epistemic practice and hence no common professional language, proponents of external solutions need to connect with potential strategic partners on a different level, i.e. strategic marketing and business development. For a company which is still in the process of transforming into a true commercial venture, this is a difficult challenge. When corporations close down whole research departments and outsource the respective competences as a matter of company strategy or shareholder satisfaction, as they do, they become both less accessible to technology-oriented founders and less predictable regarding their boundaries and possible interfaces to the outside world.

I mean two years ago, Astra Zeneca had 200 people in structural analysis. And then they applied the red pen, said it doesn't pay any more, and, as you can do in England, over the weekend 200 people stood before closed doors. (Interview 5–2)

In some cases, previous contact with the technology in question made potential strategic partners more accessible. In case 3, GENE FUNCTION, the experimental technological assemblage, which had been created in the Max Planck Institute, contained an element (gas chromatography as a method of observing cellular metabolisms), which had originally been brought in by researchers from the same chemicals company which later acted as a strategic investor. The industrial researchers had worked with this method and had suggested it during a routine on-site visit. While the respondents in this case stressed that this contact and the later creation of a strategic partnership were independent events, they also conceded that the acquaintance with the technology made it easier for the investor side to evaluate the proposal. Likewise, in case 5 AUTOIMMUNE, a scientist stressed that the very early strategic investment by an Italian corporate VC fund was possible, because representatives of the investor side were able to engage with the technology.

In addition to previous experience with particular methods, the question of schools of thought and of boundaries between epistemic sub-communities seems to be relevant. Corporate R&D executives typically described their companies' fields of specialization with reference to certain types of compounds ('small molecules', 'organic molecules') as well as scientific fields ('neurology', 'immunology') in addition to the more commercial delineations of business models (e.g. mass market vs. niche market). Yet, just as in public research funding, certain ideas and their proponents seem to be able to temporarily assume positions of dominance in organizations. In case 2 SYNTHESIS, for example, respondents repeatedly pointed out that the field of peptide chemistry is subject to very volatile levels of popularity in corporate R&D departments. In case 8 NEURON, the key entrepreneur explained that for roughly the past decade all major pharmaceutical companies who researched drug candidates for Alzheimer's disease focused on an immunological approach which put the occurrence of certain protein plaques in the brain in the centre of attention. Alternative approaches therefore received little attention.

Therefore, interesting and enrolling a strategic investor requires the identification of very time-sensitive opportunity windows, in which potential partners are willing to open themselves up to external technology proposals. In the situation, both, the business development interests and the internally pursued techno-scientific approaches, ideas and evaluative principles need to be aligned under a shared perception of value vis-à-vis a potential biotechnological partner company. Thus, the creation of 'interest proximity' (Ibert and Müller 2015) is extremely sensitive to both temporary opportunities and the social skill of actors in identifying, co-creating and exploiting them. Sharing an epistemic enculturation and possibly experience with as well as preferences for particular technologies with the future partners is seen as the most promising point of entry. Hence, success is very much a question of 'being there', of establishing and maintaining contacts and participating in the same events. One respondent, a 'pipeline builder', went so far as to claim that an established personal acquaintance was indispensable.

The cases show a great variety of pathways and material, time-spatial configurations in which a strategic relation can play out in practice. In this respect, too, strategic investment is more versatile than purely financial investment. Furthermore, the relative contribution of a strategic relation to the overall course of an innovation process differs from case to case. Strategic relations interact with other investment relationships, either laterally or sequentially. Some strategic relations are exclusive and prove permanent. Others are only temporary and 'fail' in their original purpose to produce an envisioned market entry, but still have an effect. In the latter case, for biotechnology entrepreneurs, disentangling or freeing themselves from the bindings of a strategic relation is both a challenge and an opportunity. Again, pre-existing industrial relations are resources in the necessary relational work, just as they are in creating a strategic relation.

Instead of framing strategic investment as the opposite of VC investment, both logics can also be brought in conjunction. This approach is actively pursued by corporations, who strive to appropriate the advantages of venture capitalism while retaining a strategic perspective. To strategic investors, venture finance is an additional source of funding, which allows them to participate in innovation processes without being solely responsible for funding them. It also affords the opportunity to observe developments and engage with the most promising and fitting ones without tying oneself to them immediately. The way in which this approach is put into practice is the setting up of corporate VC funds. Corporations assign funds in the normal size of a VC fund (an amount in the high double digit or low triple digit millions range) to investment. These funds are managed by a separate management team which is organizationally affiliated with the corporation, but enjoys some freedom to invest. Crucially, these funds mimic the orders of venture capitalism; they strive to yield profits in a similar magnitude, and also have similar investment time frames, although none of these constraints are 'factually' necessary. There are no pension funds who demand a specified return after a specified time period. Instead, the purpose of this mimicry is the ability to co-invest with purebred VC funds, i.e. to be accepted by them.

One respondent, a corporate VC manager, explained that the business model is similar to purebred VC, yet slightly different in practice. Corporate VC funds are willing to invest in slightly earlier and riskier developments, as they see an added benefit in knowing a technology early. They are also slightly more patient and can sustain an investment relation longer then a classic VC fund. Crucially, in both initial assessment and in playing their role as an investor, they can draw on corporate R&D, product development and production capabilities as well as experience with regulatory processes. As long as the chance for a profitable exit is not reduced, VC investors welcome this added value of corporate venture capitalists. As a consequence, should the corporation behind the fund decide to acquire a technology (by licensing it or by acquiring the investee company), it has to pay a market price. Corporate venture funds are thus a tool to create commonality with the VC scene and to scan for new technologies while avoiding too-early or too-binding entanglements with technology driven founders. In case 5 AUTOIMMUNE, all these factors were present and enabled an early stage investment which was crucial for the entire innovation process, including the possibility of opening the investor consortium to regular VCs.

There is a surprising diversity of valuation logics in corporate venture investment. Specifically, there are diverse ways in which industrial product development relates to the selection of investee companies. Four approaches could be discerned: Corporate venture funds can be closely connected to internal product development efforts. In this case, fund managers search for investee firms in technology and application fields which are perceived as similar to the company's own activities (for example, immunology). In this arrangement, company internal R&D scientists are closely linked to the evaluation of investment alternatives. They also can initiate searches for potential investment targets when they identify a specific desideratum. Corporate venture funds can also be largely indifferent to the mother company's product development efforts and concentrate on the financial aspect. In this case, strategic benefits from the engagement with venture capitalism consist solely of the extension of the company's sphere of contact and influence. Corporate venture funds can be used as scouting devices and as tools to establish a presence in a market, which is considered attractive. For example, in the 1990s, German corporations began establishing VC funds in the United States in order to benefit from the country's research and start-up landscape. Finally, corporate venture funds can be used to counter blind spots and scan for new developments, which would be overlooked if investment were dominated by an internal product development logic. In this case, corporate venture funds invest specifically in technologies and approaches which are not pursued internally. Hence, relational distance in investment relationships is created as a means to organize surprises.

Although regular venture capitalists appreciate the industrial expertise delivered by corporate VC funds, they also perceive disadvantages of an association of VC and corporate strategy. One interviewed VC manager pointed out that corporate venture funds are as much subject to unforeseeable shifts in company strategy as all other corporate innovation activities.

In some cases, most notably in case 8 NEURON, strategic investors outside any established market for biotechnology products used their investments to 'buy into' this field – for instance generics companies and non-pharmaceutical production companies. As in the case of newly emerging VC business models, such shifts rely on opportunities which are highly sensitive to time and space.

As a result, existent value orders are recombined in an innovative way and new circuits of capital are created. In the cases studied here, strategic investment proved indispensable: for financial reasons, but more importantly because it created the framing conditions for VC investment, which in turn enabled independent commercial ventures.

Notes

1 Temperature-based sterilization.
2 Since 2011 the *Gesetz zur Neuordnung des Arzneimittelmarktes (AMNOG)* or 'law on the reordering of the drug market' forces pharmaceutical companies a) to demonstrate a new drug's therapeutic value added vis-à-vis existing products and b) to negotiate a price with health insurance providers within one year after admission.
3 These legal changes are: the Orphan Drug Act of 1983 in the US and Regulation (EC) No 141/2000 in the EU.
4 www.alzheimers.org.uk/liraglutide; accessed 27 September 2015.
5 In relative terms. In 2014, the biggest German pharmaceutical company, Bayer, ranked 16th based on global sales (www.pmlive.com/top_pharma_list/global_reven ues). In 2013, German generics company Stada ranked 6th based on the same metric (http://de.wikipedia.org/wiki/Generikum#cite_note-40; both sources accessed 12 May 2015).
6 The term technology platform is widely used to describe both a business model and a socio-technical assemblage which is not centred on a single project. Instead, a platform is capable of performing a specific operation on a high scale and in predictable quality. Hence, it can be used to support multiple projects by, for example, creating tailored molecules or analysing functions.

5 Conclusions

Throughout the past chapters I have tried to give accounts of innovation processes in German biotechnology which provide answers to the following questions: How do these processes unfold through time and space? Through which forms of agency, understood as relational work by individuals embedded in situated social contexts, are relationships created, transformed, ended and replaced? How do these activities advance innovations? My emphasis in this book was to understand the role of investment in these processes. Investment, rather than 'finance', was understood as a type of relational work in which funds are allocated to innovation activities in the expectation of returns – financial and otherwise – based on a particular logic of appreciating and assessing the value of innovations. While venture capital (VC) is often treated as the preeminent form of financing in biotechnology, and while in the context of political economy VC is seen as a manifestation of financialization in the field of innovation, I approached investment in a more open framework, accounting for the possibility of multiple orders of worth to be enacted in investment.

I placed this work in the literature on relational economic geographies of innovation processes. Relational economic geography is not concerned with describing or analysing regional and structural formations as objects, but rather with understanding how economic actors position themselves vis-à-vis such entities, how they make use of proximity as well as distance, co-presence as well as mobility. Regarding innovation processes, I used an innovation phase model which I helped develop (Ibert and Müller 2015) as a conceptual starting point. This model contains several crucial considerations: Throughout innovation processes, relational constellations succeed each other. Each contributes in specific ways to the advancement of innovation processes, for instance by fostering the creation of ideas or by providing a context for validation. Each of these constellations or relationships is characterized by specific combinations of relational proximity and relational distance, and enacted in physical space through situated practices, co-location, co-presence and mobility.

Conceptually, I seek to add several elements to this literature, the most important one being a shift from describing relationships as entities towards addressing the work of building relationships, giving them a shared purpose, using them, changing them, ending and replacing them, as well as constructing the proximities and distances which characterize them. To this end I applied a heuristic bridge between the concept of 'orders of worth' (a conceptualization of value combining its moral and instrumental dimensions as well as its capability to assign meaning and desirability to actions, Stark 2009) and the idea of 'translations': a poststructuralist concept describing how actors struggle to enrol other actors into their agenda, wrestling them loose from competing ones. Rather than being determined by their respective context, actors can interpret and recombine the orders in which they are embedded, justifying a wide range of translation activities. I used concepts from network research, such as 'brokerage' and 'multiple insiders' to further qualify relational work. In conceptualizing actors and agency in this way I partially take up Gailing's and Ibert's (2016) notion of 'key players' in spatial development. In this conceptualization of agency, some key individuals exert influence not by virtue of their objectively given character properties, but through partially active and partially passive positionings in relational webs. Furthermore, such actors use space – opportunity structures unevenly distributed in space – as a resource for action, while also pursuing spatial agendas.

This framework, I think, is suited to help mediate between different literatures which also deal with the unfolding of innovations, but stand disconnected to process-centric approaches in relational economic geography. According to Birch (2016) there is a fundamental tension between a techno-scientific perspective on biotechnology and investment, rooted in flat ontologies, particularly Science and Technology Studies (STS) but also Social Studies of Finance, and a political economic view highlighting the structural forces of capitalism. He places his own work on 'commodification, financialization and assetization' in the context of the latter. By viewing relational work in innovation processes as a constant struggle to align actors (human as well as non-human), practices and knowledge domains under a shared purpose, the study of innovation processes can make use of key ideas from STS and Actor-Network Theory (ANT). Essentially, innovation processes are dynamics of assemblage (Müller 2015b), creating new socially pervasive formations such as, for instance, a new type of cancer therapy. On the other hand, key ideas from political economy, such as the growing trend towards financialization (i.e. a growing influence of capital markets in all areas of economic activity), power imbalances, spatially uneven development, recurring financial crises and the role of speculation can be reflected by observing how actors align themselves with orders of worth, and how objects as well as people circulate in spatially uneven and highly fluid landscapes.

Investors and the influence they exert were not treated as an external force, but rather as a crucial element in the unfolding relational dynamics. In this study, I treated the importance of finance as an open question. Given the central role VC is assigned both as a manifestation of a financialized economy and as a key financing source for biotechnology, I drew on the literature on the geography of VC as a first step in linking innovation and investment. Three key developments in this literature were highlighted: Firstly, VC, while often described as highly distance sensitive, is increasingly recognized as a form of financial intermediation (i.e. relational work) over distance, both physical and relational, involving collaborative organization (syndication) and brokerage. Secondly, in discussions whether VC activity in a region is more a supply-side or more a demand-side effect, a more relationally and ecologically oriented perspective has taken hold, highlighting relational work both across regional borders and the supply/demand divide. VC investors take part in the creation of investment opportunities. In doing so, they are highly sensitive to the relational and spatial context they are embedded in. Consequently, being located in an entrepreneurial region increases the chances of successful investment over distance. And thirdly, there is a diversity of investment logics including CVC, public VC and private VC/business angel investment, which supplement classic VC and interact with it. These interactions appear to reinforce spatial unevenness rather than countering it. However, the interactions of these different forms over time and the qualitative impact they have on innovation processes has not been studied. Furthermore, important investment forms, such as direct investment by industry corporations has not been studied in this way either.

Methodologically, I built on the recent development of innovation biographies (Butzin and Widmaier 2016) in which innovation processes are reconstructed qualitatively and in an ex post perspective. I extended the scope of this method in two ways. Firstly, I specifically included the relational work by investors. Secondly, I specifically included the linkage between relational work in innovations and relational work in the context of innovations, thereby reflecting that context conditions can change throughout the unfolding of an innovation (particularly during very drawn-out processes as they occur in biotechnology) and can be changed by actors in innovation processes. In total I studied eight innovation processes in the time span roughly between 1990 and 2015.

In Chapter 2, I discussed the situation of capital market financing for biotechnology in Germany as well as its change over time. Biotechnology in Germany grew as result of a number of intensive policy efforts and institutional changes necessitated by the fact that Germany's coordinated and manufacturing-centric economy was increasingly lagging behind in high-technology performance in the 1980s and 1990s. These include the establishment of gene centres, the BioRegio competition, a number of very generous, easily accessible early-stage financing tools and the establishment of a dedicated start-up segment in Germany's largest stock exchange. After the crisis of 2000/2001, which turned out extremely costly to the public sector and its financing institutions, public VC policy was remodelled. New mechanisms were set up, now seeking to mimic classic VC funds (building relational proximity) in order to attract such via co-financing. A division of labour was created between High-Tech Gründerfonds HTGF, a federal level seed-fund, later-stage public VC funds on state level and a purely passive co-financing mechanism provided by KfW (Reconstruction Loan Corporation).

The intermediate phase between the two latest financial crises saw a long dearth in VC financing and particularly initial public offerings (IPOs) along with a severe backlash to 'risky' business models such as pharmaceutical product development. However, the intermediate phase also saw the creation of two singular biotechnology-oriented investment offices, which went on to dominate biotechnology financing in Germany for years to come. Here, I denominated them SOUTHWEST FAMILY VC (a family VC office created with the proceeds from the sale of a generics company to a pharmaceutical corporation) and RETAIL VC (a longer-term oriented VC fund financed through private household investments collected via a retail finance network). These two players compensated, to a degree, the lack of conventional VC – in terms of direct financing, but also in terms of their capability to collaborate with international investors and thus attract international money. However, due to their specific, idiosyncratic logics and relational strategies, they appeared unattractive as syndication partners to some notable public financing actors.

Public VC, on the other hand, became more and more entangled with mid-sized VC funds focusing, among other things, on life science, but not on its very expensive and risky elements, such as bio-pharmaceutical product development. Investments by very rich individuals as well as public banks became an important financing source for conventional, small or mid-sized VC funds operating in the life sciences in Germany (rather than banks and insurance companies). Hence, two highly territorialized 'worlds' of specifically German VC emerged largely separate from each other, but also relationally distant from conventional, Anglo-American VC. The crisis of 2008/2009 aggravated this rift with public financing institutions developing a 'Lehman trauma' and banks and insurance companies turning away from VC (partially due to stricter regulation). Only very recently conventional VC financing in biotechnology has returned to relevant levels and the IPO window has opened once more, albeit on a moderate level. Yet, as I write these lines, the next crisis might be imminent.

This volatility was felt strongly by the innovators I interviewed. To the innovation processes studied here, the changing tides of biotechnology financing provided opportunities and constraints. Indeed, the unfolding of innovations was intricately linked to the capital market conditions during important junctures in innovation processes. One innovation (case 4 BIOMARKER) received its initial financing during an unusually optimistic period (the late 1990s), and this momentum pushed the innovation forward on a path which was out of reach for later innovations. This momentum did not consist of money alone, but of relational constellations unfolding in a path-dependent manner. In other cases, innovations were forced towards reorientation during phases of crisis. In case 5 AUTOIMMUNE, the dramatic lack of international conventional VC around the crisis of 2008 led to an engagement with RETAIL VC, which in turn allowed the respective company to pursue a comparatively long-term oriented strategy.

Despite the sensitivity to temporal context, the dynamics of relational work which bring about innovation processes and drive them forward showed some typical patterns of relational work in co-creating and exploiting opportunities as well as in creating, transforming, ending and replacing relationships, including investment-related ones. I discussed them under the broad categories of 'early stages' and 'later stages', recognizing of course the coarseness of this distinction. 'Early stages' covers the gradual development of a key idea, its first formulation as a commercialization project as well as its initial socio-technical manifestation as a biotechnology start-up which typically operates in a rather science-driven, experimental way and quite likely lacks a clear business strategy. 'Later stages' covers the transition towards a clear-cut product or service as well as the entry into a market – a transition which requires a profound change in knowledge practices and inter-site collaboration and often entails changes in leadership and staff.

As other studies have shown, the inception of an innovative idea is a longer-term process. It involves both exchanges across epistemic boundaries focusing on emergent topics ('conversations', Rutten 2017; 'immersion', Wenger 1998) as well as a continuous experience of unsatisfying approaches to practical problems (Ibert and Müller 2015). I found that in every studied innovation, concrete individuals were in positions to experience these dynamics and were able to use them to come up with an idea which would define an innovation. I named this form of relational work 'enthusiasm-led entrepreneurship' and the individuals pursuing it 'enthusiastic founders'. These individuals typically had a strong sense of purpose and of serving the public good by coming up with novel solutions to practical problems. They also cherished working across the boundaries of epistemic communities and engaging with users to gain inspiration. They identified strongly with their respective projects and sought to be personally associated with them – a circumstance which often led to conflict later. Thus, in terms of 'orders of worth', they combined the 'civic world', the 'world of fame' and possibly also the 'inspired world' in one role performance. The 'market order' was typically perceived as a practical necessity rather than something the actors had truly internalized.

The work was conducted in a research environment, and in most cases public research facilities. Yet, linkages to practical application were present in all cases, albeit in a variety of organizational and inter-organizational forms, thus allowing inspiration by more applied problems. In all but one case these (partial) investors founded companies, built local and trans-local networks and enrolled support including seed investors. The emergent networks and sites of practice – with a start-up company as nucleus – were highly context specific in spatial and relational terms and included ties to the academic mother institution ('institutional hybridity').

The relational embeddedness which allowed entrepreneurs, through multiple interactions over time, to identify a concrete commercialization project, only occurred through a second type of agency which I labelled 'place shaping'. The term denotes the creation of highly localized organizational and institutional environments, in which the said interactive dynamics could unfold. These were typically leading research institutions – often Max Planck Institutes or university hospitals – but also more commercially oriented environments such as an actively managed technology park. In such environments, representatives of different epistemic practices and communities could meet and work on joint projects in a collaborative fashion. For such environments to emerge, leadership was necessary. For instance, in case 3 GENE FUNCTION, a new institute had just been established which focused on a new and very disputed linkage between genetics and biochemistry. In addition, to be able to engage with users and to envision commercialization, entrepreneurs required an environment which encouraged such endeavours. Thus, in most cases distinct individuals who worked to create such environments and who acted as mentors to the respective entrepreneurs, i.e. they acted in highly individual, dyadic relationships with them, were identifiable. Often entrepreneurs, like scientists working in contested and innovative fields in general, required protection from proponents of more orthodox approaches as well as institutional legitimacy. Entrepreneurs also required the reputation and credibility of experienced scientific leaders to enrol supporters such as seed investors.

'Place shapers' were therefore identified as the second defining type of key individual actor. Through mentorship and guidance, they exerted a crucial influence on the creation of early, semi-commercial materializations of innovative ideas. These individuals typically had mixed careers combining senior positions in academia with advanced careers in more commercial fields, such as industry, business support and start-up entrepreneurship. In terms of valuation, the appreciation of border-crossing activity (both interdisciplinarity and entrepreneurship) was a means of importing new impulses into their domain, thus cultivating a reputation of innovativeness, and of serving the public good ('civic world'). Often place shapers brought ideas and preferences regarding commercialization paths into commercialization projects. Early stage investment therefore appeared as a part of the place shaping 'landscape' which shaped innovations in their early stages. In terms of financing models, these investments were tremendously diverse, including angel investment, public and semi-public seed investments, industrial and corporate venture capital (CVC) investments and, in one case, conventional VC. However, they shared two key characteristics: A highly time situation-specific interest resulting from a particular strategic orientation, and in all but one case a pre-existent linkage with the relational nexus from which the respective commercialization project emerged.

To become commercially successful, the emergent innovation relations need to transform fundamentally: Much more standardized, bureaucratized knowledge practices replace experimental ones. Formalized relations with fixed roles replace informal, enthusiasm-driven arrangements. Sites of practice become more specialized and are integrated into far-reaching trans-local networks ('pipelines'). Typically, both the locus of strategic control and of financial gains appropriation shifts in space. To achieve this transformation, a third type of relational work is required which I denominated 'pipeline building'. While this agency is distributed across a large number of actors and often a corporate form of organization, again key individuals appear as drivers and leaders. These 'pipeline builders', once more, are typically individuals with senior positions in the business world, but also with experience in and appreciation for academia and/or start-up entrepreneurship. They bring a new order – the 'industrial world' – into innovation processes, focusing on formalizing knowledge work and prioritizing projects. Crucially, large-scale VC investments in most cases occurred only after the aforementioned changes were initiated. Earlier investors often find themselves trapped in a contradiction of interests: To make their investment commercially viable, they must strive to initiate a transformation, often by searching for a new CEO. This transformation, however, is likely to lead to a loss of influence and watering down of their investment.

These findings first and foremost show that focusing on different forms of relational work throughout the time-spatial unfolding of innovations is a worthwhile endeavour. It produces relevant results. In particular, it serves to show the important role played by the agency of key individuals. Yet, it also shows how strongly such individuals' ability to act depends on their movements in time and space within and between places and territories, and thus how relational work is grounded in time and space. In such a contemporary reading of time geography (Bathelt and Schuldt 2010; Törnqvist 2004) the creative and entrepreneurial power of places becomes easier to grasp. In addition, the dichotomy of 'buzz and pipelines' (Bathelt, Malmberg and Maskell 2004) or the 'localizing' and 'globalizing' of knowledge (Birch 2012) becomes less absolute. Instead, the continuous bridging, but also re-establishing of boundaries by individuals and their biographical mobility becomes visible. The evolution of contexts or ecologies of innovations and the innovations themselves can be better grasped in their interactive dynamics – particularly if the focus on relational work is paired with an ANT/STS-inspired take on materiality.

In addition, the ability of a process-centric relational geography of innovation to mediate between more materialistic and more political economic approaches could be proven. In a political economic perspective, the discrepancy between capital allocated to biotechnology and the sector's economic returns must lead to the conclusion that biotech is almost entirely a speculative scheme. However, part of this perception is due to a firm-level focus which understands biotechnology as a type of company. By focusing on the entire innovation process, the entanglements which evade a focus on firms and everything which can be counted at firm level, become visible, as does the often quite temporary but still crucial role of the biotechnology company. STS approaches, on the other hand, were criticized for being too focused on the laboratory context. This shortcoming, too, is remedied by a process perspective.

However, a tension became visible between the analytical framework and the 'story' told by the data material. In this study, I applied a linear idea of innovation, following its 'defence' by Balconi, Brusoni and Orsenigo (2010). Indeed, innovations undergo developmental steps and transformations which cannot be reversed. Despite a general trend towards more circular and feedbacked understanding of innovation, once *one* innovation is identified and defined, it can be tracked back through its contexts, and both development phases and irreversible leaps or transitions will become visible. However, I placed the origin of innovations in scientific contexts and ascribed – to a large part – scientifically defined ideas the role of defining the core of an innovation. Hence, I followed the concept of the 'idea-centric network analysis' (Ibert and Müller 2015).

This approach could be scrutinized critically for reproducing objectivist notions of knowledge and innovation. Certainly, biotechnology is highly 'science-intensive', and it would not qualify as a case of user-driven innovations (as some innovations in medical technology do). However, what the data showed is that the creation of market opportunities – the phasing-out of a patent, a reorientation by a major player in the market who then exerts demand for a new scientific approach – is as crucial to an innovation as the creation of scientific opportunities. A more realistic perspective would thus be one which takes into view the various emergent opportunity structures which are made to gradually overlap until a match is found. With the notion of 'pipeline building' I suggested a type of relational work which achieves this match. Likely, opportunities on the market side and opportunities on the science side will be the important ones. This approach has the additional charm of being able to integrate 'macro' dynamics on both sides, such as economic boom and crisis cycles and 'fashions' in science funding.

On the market side, an interesting shift can be observed: During times of financial crisis, the only realistic option of an exit by VC investors is a sale (in some form) to industrial corporations. Hence, the market and product strategies of corporations are the arena in which opportunities emerge. In times of economic growth and capital market enthusiasm, however, innovations can be refinanced by an IPO or generally via the stock market. Thus, the locus of opportunity creation shifts between these two poles. In both cases, regulators are important co-producers (or preventers) of innovation opportunities. The fact that strategic investment by industrial corporations was more effective in bringing about meaningful development steps in the innovations I studied, compared to VC, may be due to the circumstance that crises and after-crisis periods were so prominent in the sample.

The promise of an agency-centric approach to innovation processes, as developed here, is furthermore not the only the influence of changing context conditions on innovations that can be observed, but also the impact of innovations on their context conditions. According to Mattson (2009), for instance, innovations are 'unruly': They are complex, their unfolding in time and space is unforeseeable and they can be destructive to the institutional contexts from which they emerged. The counter-idea would be one of innovations which reinforce the conditions from which they emerged. One such notion is that of 'normal innovation'. The concept emerged in debates of 'sustainability transitions' – a body of literature devoted to identifying ways towards a more sustainable energy and mobility system (Smith, Voß and Grin 2010; Truffer 2012). Innovations are considered 'normal' if they do not lead to an alteration of the fundamentally stable relationship between fossil-resource consumption and wealth (Nordhaus 2011). Depending on the way context conditions are conceptualized, findings will be differentiated. Most innovations analysed in this study would have to be qualified as 'normal' innovations. Cases 3 GENE FUNCTION and 4 BIOMARKER, once again, emerge as exceptions: They contradicted the hierarchy between Germany and the US in terms of the appropriation of financial gains from innovations and in terms of spatial centrality in knowledge commercialization.

In this study I addressed dynamics of knowledge commercialization under conditions of assetization and, more broadly, financialization. Hence, the most apparent contribution is in the field of geographies of innovation processes. Yet, there are also consequences for the study of investment dynamics in financial geography. This study addresses neither the political economy of the financial centre and financial institution, nor the assemblage of capital markets explicitly. Instead, it focuses on different forms of investment (capital marked-based and otherwise) 'in action', interacting with each other and with dynamics of knowledge production in innovation processes. Through this lens, a surprising diversity of logics of value appreciation in investment became visible. Some unexpected investors include traditional 'Mittelstand' companies seeking to branch out into new markets and new forms of VC (e.g. RETAIL VC). In addition, the temporal dynamics of investment activities was more complex than the often-cited succession of friends and family investment, business angels, VC and finally stock market-based financing.

Different metrics of value are continuously recombined and circuits of capital are continuously rewired. Appreciations of financial value are often embedded in other, non-financial or even non-economic notions of value. In this way, cultural spaces of investment are enacted on different scales – transnational, national, regional and local. Periods of crisis, of stabilization and of boom each provide specific opportunities for such re-formations. This dynamic landscape was unveiled in part because the analysis focus was on investment in action rather than financial institutions. Possibly, the study of financial and financial centres institutions in a structure-oriented framework serves to reproduce restrictive assumptions of what investment is and can be, even if it explicitly focuses on change.

Several findings in this relate to geographies of VC. The role of conventional VC, as the most frequently discussed modus of innovation financing, shifted over time. In one innovation process, which started in the 1990s, initial financing by a German VC fund laid the foundation for the creation of an independent company with a product-centric business model. In the intermediate period between 2000/2001 and 2008/2009, and more so after the last financial crisis, conventional VC funds pursued a much more cautious approach: They approached investee companies with a much stricter logic of assetization. Individual assets, such as product development projects and platform capabilities, were valued and evaluated separately, endowed with separate market strategies (such as up-scaling and qualitative upgrading for platforms; market niche strategies for products) and often sold separately. Thus, innovations with a strong influence of conventional VC mostly unfolded along decidedly undisruptive paths and catered to the needs of large corporations. Public venture funds acted as co-investors in the respective cases, thus reinforcing the dominant logic of assetization. Using VC to build an independent and 'organic' company around a particular technology appears as an unrealistic endeavour, given this trend. In unconventional VC models, however, such a strategy of company building was actively pursued. In case 2 SYNTHESIS, a platform technology 'travelled' through the more restrictive VC-based logic of valuation before being sold and attached to a new company which was actively built by a biotechnology-oriented VC family office (SOUTHWEST FAMILY VC).

By focusing on innovation processes, both the limitations and the inter-dependence of VC become visible. Rather than studying VC as a separate entity, studying investment as an element or tool in innovation-oriented corporate strategy – in defending existent market positions as well as in building new positions and new markets – is a promising avenue for financial geography. This perspective would include CVC, collaborative relations with independent VC firms and other forms of investment. It would also need to reflect the nexus between corporations and regulators as well as the relationship between market strategy, internal research and openness to external solutions in corporations. In all likelihood, large parts of strategies will be focused on preventing innovations which would be disruptive for the respective corporation's business model, or at least on moulding in such a way that they support existing business models. By applying a relational as well as materiality-oriented perspective on innovations, the intricacies of such strategies can become visible. Methodologically, however, it would be a tremendous challenge to study an innovation process as well as the dynamics in all the named arenas. Therefore, a mixed approach combining a set of innovation process case studies with an ethnographic or reconstructive study of key decision environments during a specific time period may be the way to go.

In this study I find strong support for a relational perspective on regional VC markets (Wray 2012). The creation of investment opportunities is distributed across many actors. It concerns the demand-side as well as the supply-side of VC. Over time, individuals can change sides, thus crossing but also redrawing boundaries. Such dynamics deserve further attention. Furthermore, the ability to identify and co-create investment opportunities locally and the ability to invest across physical distances are not in contradiction to each other, but rather support each other. While this perspective has gained acceptance in research it still constitutes a challenge to the assumptions implicit in public VC policy.

In Germany, VC policy has evolved. In its earliest conception, public VC financing was entirely predicated on the assumption that nothing but a money supply shortage had to be addressed. More recently, more advanced strategies of creating relational proximity were established. State and federal financing institutions (such as Land VC funds and KfW, see Chapter 2) primarily act as co-investors to conventional VC funds, exerting varying degrees of strategic influence. In the light of my findings, however, I find it highly problematic for public financing institutions to replicate the logic of conventional VC. HTGF (see Chapter 2) seeks to build a different kind of relational proximity by connecting technology-centric seed financing in Germany to the market strategies of large German corporations. Yet, this approach, too, is clearly anti-disruptive and serves to reinforce existing structures rather than allowing new industries to emerge. Given the pressing global challenges faced by industrialized societies today, public policy for innovation financing should adopt a friendlier stance towards disruption as well as innovation paths enacting diverse notions of value.

References

Adleberger, Karen (1999) A developmental German state? Explaining growth in German biotechnology and venture capital. Berkeley Roundtable on the International Economy, University of California, Berkeley, CA (*BRIE Working Papers*, 134).

Aitken, Stuart C. (2010) 'Throwntogetherness': Encounters with difference and diversity. In Dydia DeLyser, Steve Herbert, Stuart Aitken, Mike Crang, Linda McDowell (Eds.): *The SAGE Handbook of Qualitative Geography*. Los Angeles, CA: Sage, pp. 46–68.

Akrich, Madeleine; Callon, Michel; Latour, Bruno; Monaghan, Adrian (2002) The key to success in innovation part I: The art of interessement. *International Journal of Innovation Management* 6(2): 14–29.

Amin, Ash; Cohendet, Patrick (2004) *Architectures of Knowledge. Firms, Capabilities and Communities*. Oxford and New York: Oxford University Press.

Amin, Ash; Roberts, Joanne (2008) Knowing in action: Beyond communities of practice. *Research Policy* 37(2): 353–369.

Andersson, Tord; Gleadle, Pauline; Haslam, Colin; Tsitsianis, Nick (2010) Bio-pharma: A financialized business model. *Critical Perspectives on Accounting* 21(7): 631–641.

Arthur, W.Brian (2009) *The Nature of Technology: What It Is and How It Evolves*. New York: Free Press.

Avdeitchikova, Sofia (2009) False expectations: Reconsidering the role of informal venture capital in closing the regional equity gap. *Entrepreneurship and Regional Development* 21(2): 99–130.

Balconi, Margherita; Brusoni, Stefano; Orsenigo, Luigi (2010) In defence of the linear model: An essay. *Research Policy* 39(1): 1–13.

Bannier, Christina E.; Grote, Michael M. (2008) Equity gap? – Which equity gap? On the financing structure of Germany's Mittelstand. Frankfurt School of Finance and Management. Frankfurt/Main (*Frankfurt School – Working Paper Series*, 106).

Baraldi, Enrico; Strömsten, Torkel (2009) Combining and controlling resources in networks – from Uppsala to Stanford, and back again: The case of a biotech innovation. *Industrial Marketing Management* 38.

Barley, Stephen R.; Tolbert, Pamela S. (1997) Institutionalization and structuration: Studying the links between action and institution. *Organization Studies* 18(1): 93–117.

Bartholomew, Susan (1997) National systems of biotechnology innovation: Complex Interdependence in the global system. *Journal of International Business Studies* 28(2): 241–266.

Bathelt, Harald; Glückler, Johannes (2003) Toward a relational economic geography. *Journal of Economic Geography* 3(2): 117–144.

Bathelt, Harald; Glückler, Johannes (2005) Resources in economic geography: From substantive concepts towards a relational perspective. *Environment and Planning A* 37(9): 1545–1563.

Bathelt, Harald; Schuldt, Nina (2010) International trade fairs and Global Buzz, Part I: Ecology of Global Buzz. *European Planning Studies* 18(12): 1957–1974.

Bathelt, Harald; Glückler, Johannes (2013) Institutional change in economic geography. *Progress in Human Geography*. https://doi.org/10.1177%2F0309132513507823.

Bathelt, Harald; Malmberg, Anders; Maskell, Peter (2004) Clusters and knowledge: Local buzz, global pipelines and the process of knowledge creation. *Progress in Human Geography* 28(1): 31–56.

Battilana, Julie; Leca, Bernard; Boxenbaum, Eva (2009) How actors change institutions: Towards a theory of institutional entrepreneurship. *The Academy of Management Annals* 3(1): 65–107.

Berglund, Henrik; Hellström, Tomas; Sjölander, Sören (2007) Entrepreneurial learning and the role of venture capitalists. *Venture Capital* 9(3): 165–181.

Bindseil, Kai Uwe (2005) Wie kann sich Berlin zu einem international führenden Biotechnologiecluster entwickeln? In Benjamin-Immanuel Hoff, Harald Wolf (Eds.): *Berlin – Innovationen für den Sanierungsfall*. Wiesbaden: VS Verlag, pp. 150–165.

Birch, Kean (2012) Knowledge, place, and power: Geographies of value in the bioeconomy. *New Genetics and Society. Critical Studies of Contemporary Biosciences* 31(2): 183–201.

Birch, Kean (2016) Rethinking value in the bio-economy: Finance, assetization, and the management of value. *Science, Technology and Human Values* 42(3): 460–490.

Block, Joern; Sandner, Philipp (2009) What is the effect of the financial crisis on venture capital financing? Empirical evidence from US Internet start-ups. *Venture Capital* 11(4): 295–309.

Boltanski, Luc; Thévenot, Laurent (2006) *On Justification: Economies of Worth*. Princeton, NJ: Princeton University Press.

Boltanski, Luc; Chiapello, Eve (2007) *The New Spirit of Capitalism*. London and New York: Verso.

Boschma, Ron (2005) Proximity and innovation: A critical assessment. *Regional Studies* 39(1): 61–74.

Breznitz, Dan (2007) *Innovation and the State. Political Choice and Strategies for Growth in Israel, Taiwan and Ireland*. New Haven, CT and London: Yale University Press.

Brinks, Verena; Ibert, Oliver (2015) Mushrooming entrepreneurship: The dynamic geography of enthusiast-driven innovation. *Geoforum* 65: 363–373.

Brown, John Seely; Duguid, Paul (2001) Knowledge and organization: A social-practice perspective. *Organization Science* 12(2): 198–213.

Bruns, Elke; Köppel, Johann; Ohlhorst, Dörte; Schön, Susanne (2008) *Die Innovationsbiographie der Windenergie. Absichten und Wirkungen von Steuerungsimpulsen*. Berlin: LIT-Verlag (Innovationsforschung, 2).

Burt, Ronald S. (2004) Structural holes and good ideas. *American Journal of Sociology* 110(2): 349–399.

Butzin, Anna; Rehfeld, Dieter (2008) *Innovationsbiographien in der Bauwirtschaft*. Edited by Bundesamt für Bauwesen und Raumordnung.

Butzin, Anna; Rehfeld, Dieter (2012) Forschungs - und Entwicklungsdienstleister im Innovationssystem der Nanotechnologie - drei Innovationsbiographien. In Anna Butzin, Dieter Rehfeld, Brigitta Widmaier (Eds.): *Innovationsbiographien. Räumliche und Sektorale Dynamik*. Baden-Baden: Nomos, pp. 139–156.

Butzin, Anna; Widmaier, Bridgitter (2016) Exploring territorial knowledge dynamics through innovation biographies. *Regional Studies* 50(2): 220–232.

Callon, Michel (1998) Introduction: The embeddedness of economic markets in economics. *The Sociological Review* 46(S1): 1–57.

Callon, Michel (2007) Some elements of a sociology of translation: Domestication of the scallops and the fishermen of St. Brieuc Bay. In Kristin Asdal, Brita Brenna, Ingunn Moser (Eds.): *Technoscience: The Politics of Interventions*. Oslo: Unipub, pp. 57–78.

Callon, Michel (2008) Economic markets and the rise of interactive agencements: From prosthetic agencies to habilitated agencies. In Trevor Pinch, Richard Swedberg (Eds.): *Living in a Material World: Economic Sociology Meets Science and Technology Studies*. Cambridge, MA and London: MIT Press.

Callon, Michel; Latour, Bruno (1981) Unscrewing the big Leviathan: How actors macro-structure reality and how sociologists help them to do so. In Karin Knorr Cetina, Aaron V. Cicourel (Eds.): *Advances in Social Theory and Methodology: Towards an Integration of Micro- and Macro-Sociologies*. Boston, MA, London and Henley: Routledge and Kegan Paul, pp. 277–303.

Callon, Michel; Muniesa, Fabian (2005) Peripheral vision: Economic markets as calculative collective devices. *Organization Studies* 26(8): 1229–1250.

Canzler, Weert; Wentland, Alexander; Simon, Dagmar (2011) Wie entstehen neue Innovationsfelder? Vergleich der Formierungs- und Formungsprozesse in der Biotechnologie und Elektromobilität. *WZB* (Discussion Papers Forschungsgruppe Wissenschaftspolitik, SP III 2011–2601).

Carmona, Matthew (2014) The place-shaping continuum: A theory of urban design process. *Journal of Urban Design* 19(1): 2–36.

Casper, Steven; Kettler, Hannah (2001) National institutional frameworks and the hybridization of entrepreneurial business models: The German and UK biotechnology sectors. *Industry and Innovation* 8(1): 5–30.

Casper, Steven; Lehrer, Mark; Soskice, David (2006) Can high-technology industries prosper in Germany? Institutional frameworks and the evolution of the German software and biotechnology industries. *Industry and Innovation* 6(1): 5–24.

Chen, Henry; Gompers, Paul A.; Kovner, Anna; Lerner, Josh (2010) Buy local? The geography of venture capital. *Journal of Urban Economics* 67(1): 90–102.

Clark, Colin (2008) The impact of entrepreneurs' oral 'pitch' presentation skills on business angels' initial screening investment decisions. *Venture Capital* 10(3): 257–279.

Coenen, Lars; Moodysson, Jerker; Asheim, Bjorn (2004) Nodes, networks and proximities: On the knowledge dynamics of the Medicon Valley biotech cluster. *European Planning Studies* 12(7): 1003–1018.

Coenen, Lars; Moodysson, Jerker; Ryan, Camille D.; Asheim, Bjorn; Philips, Peter (2006) Comparing a pharmaceutical and an agro-food bioregion: On the importance of knowledge bases for socio-spatial Patterns of innovation. *Industry and Innovation* 13(4): 393–414.

Collier, David; Mahoney, James (1996) Insights and pitfalls: Selection bias in qualitative research. *World Politics* 49(1): 56–91.

Colyvas, Jeannette A.; Powell, Walter W. (2006) Roads to institutionalization: the remaking of boundaries between public and private science. In Barry M. Staw

(Ed.): *Research in Organizational Behavior: An Annual Series of Analytical Essays and Critical Reviews*. Amsterdam: Elsevier, pp. 305–353.

Cook, Scott D. N.; Brown, John Seely (1999) Bridging epistemologies: The generative dance between organizational knowledge and organizational knowing. *Organization Science* 10(4): 381–400.

Cook, Ian; Harrison, Michelle (2007) Follow the thing: West India hot pepper sauce. *Space and Culture* 10(1): 40–63.

Cooke, Philip (2002) Biotechnology clusters as regional sectoral innovation systems. *International Regional Science Review* 25(1): 8–37.

Cooper, Melinda (2007) *Life as Surplus: Biotechnology and Capitalism in the Neoliberal Era*. Seattle, WA and London: University of Washington Press.

Cooper, Melinda (2012) The pharmacology of distributed experiment: User-generated drug innovation. *Body and Society* 18(3–4): 18–43.

Crevoisier, Oliver (2014) Beyond territorial innovation models: The pertinence of the territorial approach. *Regional Studies* 48(3): 551–561.

Crevoisier, Oliver; Jeannerat, Hugues (2009) Territorial knowledge dynamics: From the proximity paradigm to multi-location milieus. *European Planning Studies* 17(8): 1223–1241.

Crouch, Colin (2005) *Capitalist Diversity and Change: Recombinant Governance and Institutional Entrepreneurs*. New York: Oxford University Press.

Cumming, Douglas (2008) Contracts and exits in venture capital finance. *The Review of Financial Studies* 21(5): 1948–1982.

DeRuiter, Jack; Holston, Pamela S. (2012) Drug patent expirations and the 'patent cliff'. *U.S. Pharmacist* 37(6): 12–20.

Dewey, John (1933) *How We Think*. Boston, MA: Heath.

DiMaggio, Paul J.; Powell, Walter W. (1991) Introduction. In Walter W. Powell, Paul J. DiMaggio (Eds.): *The New Institutionalism in Organizational Analysis*. Chicago, IL and London: University of Chicago Press, pp. 1–38.

Djelic, Marie-Laure; Quack, Sigrid (Eds.) (2010) *Transnational Communities. Shaping Global Economic Governance*. Cambridge and New York: Cambridge University Press.

Dohse, Dirk (2000) Technology policy and the regions: The case of the BioRegio contest. *Research Policy* 29: 1111–1133.

Dosi, Giovanni; Grazzi, Marco (2010) On the nature of technologies: Knowledge, procedures, artifacts and production inputs. *Cambridge Journal of Economics* 34: 173–184.

Edmondson, Amy C. (2000) The innovation journey. *Academy of Management Review* 25(4).

Ernst & Young (2010) Neue Spielregeln. Deutscher Biotechnologie-Report 2010.

Ernst & Young (2015) Momentum nutzen. Politische Signale setzen für Eigenkapital und Innovation. Deutscher Biotechnologie-Report 2015.

Ernst & Young (2017) Spot on Innovation! Deutscher Biotechnologie-Report 2017.

Ernst & Young (2018) Sprung nach vorne! Modell Deutschland: Von der Biologie zur Innovation. 20 Jahre Deutscher Biotechnologie-Report.

Ettlinger, Nancy (2004) Toward a critical theory of untidy geographies: The spatiality of emotions in consumption and production. *Feminist Economics* 10(3): 21–54.

Faulconbridge, James (2010) Global architects: Learning and innovation through communities and constellations of practice. *Environment and Planning A* 42(12): 2842–2858.

Feldman, Maryann P.; Francis, Johanna L. (2003) Fortune favours the prepared region: The case of entrepreneurship and the Capitol Region biotechnology cluster. *European Planning Studies* 11(7): 765–788.

Ferrary, Michel; Granovetter, Mark (2009) The role of venture capital firms in Silicon Valley's complex innovation network. *Economy and Society* 38(2): 326–359.

Fetzer, Thomas (2010) Industrial democracy in the European community: Trade unions as a defensive transnational community 1968–1988. In Marie-Laure Djelic, Sigrid Quack (Eds.): *Transnational Communities. Shaping Global Economic Governance.* Cambridge and New York: Cambridge University Press, pp. 282–304.

Fligstein, Neil; McAdam, Doug (2012) *A Theory of Fields.* Oxford and New York: Oxford University Press.

Florida, Richard; Kenney, Martin (2000) Venture capital in Silicon Valley: Fueling new firm formation. In Kenney, Martin (Ed.): *Understanding Silicon Valley: The Anatomy of an Entrepreneurial Region.* Stanford, CA: Stanford University Press.

Franke, Nikolaus; Shah, Sonali (2003) How communities support innovative activities: an exploration of assistance and sharing among end-users. *Research Policy* 32: 57–178.

Freitag, Michael; Student, Dietmar (2013) Susanne Klatten. Das Coming-Out. In *Manager Magazin*, 18 June. Available online at www.manager-magazin.de/maga zin/artikel/susanne-klatten-uebernimmt-aufsichtsratsvorsitz-bei-sgl-carbon-a -905265-3.html, accessed 20 September 2015.

French, Shaun; Leyshon, Andrew; Wainwright, T. (2011) Financializing space, spacing financialization. *Progress in Human Geography* 35(6): 798–819.

Frenken, Koen; van Oort, Frank; Verburg, Thijs (2007) Related variety, unrelated variety and regional economic growth. *Regional Studies* 41(5): 685–697.

Frickel, Scott; Gross, Neil (2005) A general theory of scientific/intellectual movements. *American Sociological Review* 70(2): 204–232.

Fritsch, Michael; Schilder, Dirk (2006) Is venture capital a regional business? The role of syndication (*Freiberg Working Papers*, 9).

Fritsch, Michael; Medrano, Luis F. (2010) *The Spatial Diffusion of a Knowledge Base – Laser Technology Research in West Germany, 1960–2005* (Jena Economic Research Papers, 048).

Fuchs-Heinritz, Werner (2009) *Biographische Forschung. Eine Einführung in Praxis und Methoden.* 4. Auflage. Wiesbaden: VS Verlag für Sozialwissenschaften (Hagener Studientexte zur Soziologie).

Gailing, Ludger; Ibert, Oliver (2016) Schlüsselfiguren: Raum als Gegenstand und Ressource des Wandels. *Raumforschung und Raumordnung* 74(5): 391–403.

Gapp, Rod; Fisher, Ron (2007) Developing an intrapreneur-led three-phase model of innovation. *International Journal of Ent Behaviour and Research* 13(6): 330–348.

Garon, Sheldon (2012) *Beyond Our Means: Why America Spends While the World Saves.* Princeton, NJ: Princeton University Press.

George, Gerard; McGahan, Anita; Prabhu, Jaideep (2012) Innovation for inclusive growth: Towards a theoretical framework and a research agenda. *Journal of Management Studies* 49(4): 661–683.

Geyer, Anton; Heimer, Thomas (2010) *Evaluierung des High-Tech Gründerfonds. Endbericht. Studie im Auftrag des Bundesministeriums für Wirtschaft und Technologie.* Edited by Bundesministerium für Wirtschaft und Technologie.

Giesecke, Susanne (2000) The contrasting roles of government in the development of biotechnology industry in the US and Germany. *Research Policy* 29: 205–223.

Gilding, Michael (2008) 'The tyranny of distance': Biotechnology networks and clusters in the antipodes. *Research Policy* 37(6–7): 1132–1144.

Gläser, Jochen; Laudel, Grit (2010) *Experteninterviews und qualitative Inhaltsanalyse als Instrumente rekonstruierender Untersuchungen*. 4. Auflage. Wiesbaden: VS Verlag für Sozialwissenschaften (Lehrbuch).

Glennerster, Rachel; Kremer, Michael; Williams, Heidi (2006) Creating markets for vaccines. *Innovations. Technology Governance Globalization* 1(1): 67–79.

Gompers, Paul A. (1994) The rise and fall of venture capital. *Business and Economic History* 23(2).

Gompers, Paul A.; Lerner, Josh (2001) The venture capital revolution. *Journal of Economic Perspectives* 15(2): 145–168.

Grabher, Gernot; Ibert, Oliver (2014) Distance as asset? Knowledge collaboration in hybrid virtual communities. *Journal of Economic Geography* 14(1): 97–123.

Griffith, Terri L.; Yam, Patrick J.; Subramaniam, Suresh (2007) Silicon valley's 'one-hour' distance rule and managing return on location. *Venture Capital* 9(2): 85–106.

Gross, Matthias (2010) *Ignorance and Surprise: Science, Society, and Ecological Design*. Cambridge, MA.: MIT Press.

Hagel, John; Brown, John Seely; Davison, Lang (2010) *The Power of Pull: How Small Moves, Smartly Made, Can Set Big Things in Motion*. New York: Basic Books.

Hall, Sarah; Appleyard, Lindsey (2009) 'City of London, City of Learning'? Placing business education within the geographies of finance. *Journal of Economic Geography* 9(5): 597–617.

Hall, Peter A.; Soskice, David (Eds.) (2001) *Varieties of Capitalism. Institutional Foundations of Comparative Advantage*. New York: Oxford University Press.

Haller, Lea (2011) Angewandte Forschung? Cortison zwischen Hochschule, Industrie und Klinik. In Florian Hoof, Eva-Maria Jung, Ulrich Salaschek (Eds.): *Jenseits des Labors. Transformationen von Wissen zwischen Entstehungs- und Anwendungskontext*. Bielefeld: transcript, pp. 171–198.

Harrison, Richard T.; Mason, Colin M. (2000a) Editorial: The role of the public sector in the development of a regional venture capital industry. *Venture Capital: An International Journal of Entrepreneurial Finance* 2(4): 243–253.

Harrison, Richard T.; Mason, Colin M. (2000b) Venture capital market complementarities: The links between business angels and venture capital funds in the United Kingdom. *Venture Capital* 2(3): 223–242.

Hautala, Johanna; Jauhiainen, Jussi S. (2014) Spatio-temporal processes of knowledge creation. *Research Policy* 43(4): 655–668.

Haythornthwaite, Caroline (2001) Exploring multiplexity: Social network structures in computer-supported distance learning classes. *The Information Society* 17(21): 211–226.

Helbrecht, Ilse (2011) Die Welt als Horizont - Zur Produktion globaler Expertise in der Weltgesellschaft. In Oliver Ibert, Hans Joachim Kujath (Eds.): *Räume der Wissensarbeit. Zur Funktion von Nähe und Distanz in der Wissensökonomie*. Wiesbaden: VS Verlag für Sozialwissenschaften, pp. 103–124.

Hellmann, Thomas; Puri, Manju (2002) Venture capital and the professionalization of start-up firms: Empirical evidence. *The Journal of Finance* 57(1): 169–197.

Hochberg, Yael V.; Ljungqvist, Alexander; Lu, Yang (2007) Whom you know matters: Venture capital networks and investment performance. *The Journal of Finance* 62(1): 251–301.

Hoof, Florian; Jung, Eva-Maria; Salaschek, Ulrich (Eds.) (2011) Jenseits des Labors. Transformationen von Wissen zwischen Entstehungs- und Anwendungskontext. Bielefeld: transcript.

Howells, Jeremy (2012) The geography of knowledge: Never so close but never so far apart. *Journal of Economic Geography* 12(5): 1003–1020.

Hutter, Michael; Stark, David (2015) Pragmatist perspectives on valuation: An introduction. In Michael Hutter, David Stark (Eds.): *Moments of Valuation: Exploring Sites of Dissonance*. Oxford: Oxford University Press, pp. 1–14.

Ibert, Oliver (2007) Towards a geography of knowledge creation: The ambivalences between 'knowledge as an object' and 'knowing in practice'. *Regional Studies* 41(1): 103–114.

Ibert, Oliver (2010) Relational distance: Sociocultural and time–spatial tensions in innovation practices. *Environment and Planning A* 42(1): 187–204.

Ibert, Oliver; Thiel, Joachim (2009) Situierte Analyse, dynamische Räumlichkeiten. Ausgangspunkte, Perspektiven und Potenziale einer Zeitgeographie der wissensbasierten Ökonomie. *Zeitschrift für Wirtschaftsgeographie* 53(4): 209–223.

Ibert, Oliver; Kujath, Hans Joachim (2011) Wissensarbeit aus räumlicher Perspektive - Begriffliche grundlagen und Neuausrichtungen im Diskurs. In Oliver Ibert, Hans Joachim Kujath (Eds.): *Räume der Wissensarbeit. Zur Funktion von Nähe und Distanz in der Wissensökonomie*. Wiesbaden: VS Verlag für Sozialwissenschaften, pp. 9–48.

Ibert, Oliver; Schmidt, Suntje (2014) Once you are in you might need to get out: Adaptation and adaptability in volatile labor markets – the case of musical actors. *Social Sciences* 3(1): 1–23.

Ibert, Oliver; Müller, Felix C. (2015) Network dynamics in constellations of cultural differences: Relational distance in innovation processes in legal services and bio-technology. *Research Policy* 44(1): 181–194.

Ibert, Oliver; Müller, Felix C.; Stein, Axel (2014) Produktive Differenzen. Eine dynamische Netzwerkanalyse von Innovationsprozessen. Bielefeld: transcript.

Jarzabkowski, P.; Matthiesen, Jane; van de Ven, Andrew H. (2010) Doing which work? A practice approach to institutional pluralism. In Thomas B. Lawrence, Roy Suddaby, Bernard Leca (Eds.): *Institutional Work: Actors and Agency in Institutional Studies of Organizations*. Cambridge: Cambridge University Press, pp. 284–316.

Jeannerat, Hugues (2013) Staging experience, valuing authenticity: Towards a market perspective on territorial development. *European Urban and Regional Studies* 20(4): 370–384.

Jessop, Bob; Brenner, Neil; Jones, Martin (2008) Theorizing sociospatial relations. *Environment and Planning D Society and Space* 26: 389–401.

Johnson, Cathryn; Dowd, Timothy J.; Ridgeway, Cecilia L. (2006) Legitimacy as a social process. *Annual Review of Sociology* 32: 53–78.

Jones, Andrew (2013) Geographies of production I: Relationality revisited and the 'practice shift' in economic geography. *Progress in Human Geography* 38(4): 605–615.

Jones, Andrew; Murphy, James T. (2011) Theorizing practice in economic geography: Foundations, challenges, and possibilities. *Progress in Human Geography* 35(3): 366–392.

Kaitin, Ken I. (2010) Deconstructing the drug development process: The new face of innovation. *Clinical Pharmacology and Therapeutics* 87(3): 356–361.

Karberg, Sascha (2009) Biotech's perfect storm. *Cell* 138(3): 413–415.

Kenney, Martin (2012) Venture capital has a role, but don't forget nice-growth firms. In Ministry of Employment and the Economy, Finland (Ed.): *Kasvuyrityskatsaus 2012*, pp. 60–72.

Klagge, Britta; Martin, Ron (2005) Decentralized versus centralized financial systems: Is there a case for local capital markets? *Journal of Economic Geography* 5: 387–421.

Klagge, Britta; Peter, Carsten (2009) Wissensmanagement in Netzwerken unterschiedlicher Reichweite. Das Beispiel des Private Equity-Sektors in Deutschland. *Zeitschrift für Wirtschaftsgeographie* 53(1–2): 69–88.

Kline, Stephen J.; Rosenberg, Nathan (1986) An overview of innovation. In Ralph Landau, Nathan Rosenberg (Eds.): *The Positive Sum Strategy: Harnessing Technology for Economic Growth*. Washington DC: National Academy Press, pp. 275–305.

Knoben, Joris; Oerlemans, Leon A. G. (2006) Proximity and inter-organizational collaboration: A literature review. *International Journal of Management Reviews* 8(2): 71–89.

Knorr Cetina, Karin (1984) The fabrication of facts: Toward a microsociology of scientific knowledge. In Nico Stehr, Volker Meja (Eds.): *Society and Knowledge*. Oxford: Transaction Books, pp. 223–244.

Knorr Cetina, Karin (1999) *Epistemic Cultures: How the Sciences Make Knowledge*. Cambridge, MA; London: Harvard University Press.

Knorr Cetina, Karin (2001) Objectual practice. In Theodore R. Schatzki, Karin Knorr Cetina, E. von Savigny (Eds.): *The Practice Turn in Contemporary Theory*. London: Routledge, pp. 175–188.

Kolympiris, Christos; Kalaitzandonakes, Nicholas; Miller, Douglas (2011) Spatial collocation and venture capital in the US biotechnology industry. *Research Policy* 40(9): 1188–1199.

Kortum, Samuel; Lerner, Josh (2000) Assessing the contribution of venture capital to innovation. *The RAND Journal of Economics* 31(4): 674–692.

Kotz, David M. (2011) Financialization and neoliberalism. In Gary Teeple, Stephen McBride (Eds.): *Relations of Global Power: Neoliberal Order and Disorder*. Toronto: University of Toronto Press, pp. 1–18.

Krippner, Greta R. (2005) The financialization of the American economy. *Socio-Economic Review* 3(2): 173–208.

Kutter, Susanne (2014) Abschied von der grünen Biotechnik. Genfood: Deutschland steht sich selbst im Weg. In *Wirtschaftswoche*, 18 January. Available online at www.wiwo.de/technologie/forschung/abschied-von-der-gruenen-biotechnik-genfood-deutschland-steht-sich-selbst-im-weg/9312892.html, 1 March 2015.

Lange, Knut (2008) Institutional embeddedness and the strategic leeway of actors: The case of the German therapeutical biotech industry. *Socio-Economic Review* 7(2): 181–207.

Large, David; Muegge, Steven (2008) Venture capitalists' non-financial value-added: An evaluation of the evidence and implications for research. *Venture Capital* 10(1): 21–53.

Latour, Bruno (1994) On technical mediation. *Common Knowledge* 3(2): 29–64.

Latour, Bruno (1996) On actor-network theory: A few clarifications. *Soziale Welt* 47(4): 369–381.

Latour, Bruno (1999) *Pandora's Hope: Essays on the Reality of Science Studies: An Essay on the Reality of Science Studies*. Cambridge, MA: Harvard University Press.

Latour, Bruno (2005) *Reassembling the Social: An Introduction to Actor-Network Theory*. Oxford: Oxford University Press.

Latour, Bruno; Woolgar, Steve (1979) *Laboratory Life: The Construction of Scientific Facts*. London: Sage.

Lave, Jean; Wenger, Etienne (1991) *Situated Learning: Legitimate Peripheral Participation*. Cambridge and New York: Cambridge University Press.

Lawrence, Thomas B.; Suddaby, Roy; Leca, Bernard (Eds.) (2010) *Institutional Work. Actors and Agency in Institutional Studies of Organizations*. Cambridge: Cambridge University Press.

Lazonick, William; Tulum, Öner (2011) US biopharmaceutical finance and the sustainability of the biotech business model. *Research Policy* 40(9): 1170–1187.

Lehtonen, Oskari; Lahti, Tom (2009) The role of advisors in the venture capital investment process. *Venture Capital* 11(3): 229–254.

Lerner, Josh (1994) The syndication of venture capital investments. *Financial Management* 23(3): 16–27.

Livingstone, David N. (2003) *Putting Science in its Place: Geographies of Scientific Knowledge*. Chicago, IL: University of Chicago Press.

Lorenz, Edward (2001) Models of cognition, the contextualisation of knowledge and organisational theory. *Journal of Management and Governance* 5: 307–330.

MacKenzie, Donald (2008) *An Engine, Not A Camera: How Financial Models Shape Markets*. Cambridge, MA: MIT Press.

MacKenzie, Donald (2009) *Material Markets: How Economic Agents Are Constructed*. Oxford and New York: Oxford University Press.

Madden, Adrian (2010) The community leadership and place-shaping roles of English local government: Synergy or tension? *Public Policy and Administration* 25(2): 175–193.

Madill, Judith J.; Haines, Jr, George H.; Riding, Allan L. (2005) The role of angels in technology SMEs: A link to venture capital. *Venture Capital* 7(2): 107–129.

Marston, Sallie A.; Jones, John Paul; Woodward, Keith (2005) Human geography without scale. *Transactions of the Institute British Geographers* 30(4): 416–432.

Martin, Ron; Sunley, Peter; Turner, Dave (2002) Taking risks in regions: The geographical anatomy of Europe's emerging venture capital market. *Journal of Economic Geography* 2: 121–150.

Martin, Ron; Berndt, Christian; Klagge, Britta; Sunley, Peter (2005) Spatial proximity effects and regional equity gaps in the venture capital market: Evidence from Germany and the United Kingdom. *Environment and Planning A* 37(7): 1207–1231.

Mason, Colin M. (2009) Venture capital in crisis? *Venture Capital* 11(4): 279–285.

Mattsson, Henrik (2009) Geographies of unruly innovation. *Zeitschrift für Wirtschaftsgeographie* 53(4): 224–234.

Mazzucato, Mariana (2013) *The Entrepreneurial State: Debunking Public vs. Private Sector Myths*. London and New York: Anthem Press.

McGoey, Lindsey (2012) The logic of strategic ignorance. *The British Journal of Sociology* 63(2): 533–576.

Menzel, Max-Peter (2015) Interrelating dynamic proximities by bridging, reducing and producing distances. *Regional Studies* 49(11): 1892–1907.

Miettinen, Reijo; Virkkunen, Jaakko (2005) Epistemic objects, artefacts and organizational change. *Organization* 12(3): 437–456.

Miloud, Tarek; Aspelund, Arild; Cabrol, Mathieu (2012) Startup valuation by venture capitalists: An empirical study. *Venture Capital* 14(2–3): 151–174.

Mirowski, Philip (2012) The modern commercialization of science is a passel of Ponzi schemes. *Social Epistemology. A Journal of Knowledge, Culture and Policy* 26(3–4): 285–310.

Mittra, James (2007) Life science innovation and the restructuring of the pharmaceutical industry: Merger, acquisition and strategic alliance behaviour of large firms. *Technology Analysis and Strategic Management* 19(3): 279–301.

Montalban, Matthieu; Sakinc, Mustafa E. (2013) Financialization and productive models in the pharmaceutical industry. *Industrial and Corporate Change* 22(4): 981–1030.

Moodysson, Jerker (2008) Principles and practices of knowledge creation: On the organization of 'Buzz' and 'Pipelines' in life science communities. *Economic Geography* 84(4): 449–469.

Morgan, Kevin (2004) The exaggerated death of geography: Learning, proximity and territorial innovation systems. *Journal in Economic Geography* 4(1): 3–21.

Morgan, Glenn; Kobo, Izumi (2010) Private equity in Japan: Global financial markets and transnational communities. In Marie-Laure Djelic, Sigrid Quack (Eds.): *Transnational Communities: Shaping Global Economic Governance.* Cambridge and New York: Cambridge University Press, pp. 130–152.

Moulaert, Frank; Sekia, Farid (2003) Territorial innovation models: A critical survey. *Regional Studies* 37(3): 289–302.

Müller, Christian (2002) The evolution of the biotechnology industry in Germany. *TRENDS in Biotechnology* 20(7): 287–290.

Müller, Martin (2012) Opening the black box of the organization: Socio-material practices of geopolitical ordering. *Political Geography* 31(6): 379–388.

Müller, Martin (2015a) A half-hearted romance? A diagnosis and agenda for the relationship between economic geography and actor-network theory (ANT). *Progress in Human Geography* 39(1): 65–86.

Müller, Martin (2015b) Assemblages and actor-networks: Rethinking socio-material power, politics and space. *Geography Compass* 9(1): 27–41.

Müller, Felix C.; Ibert, Oliver (2015) (Re-)sources of innovation: Understanding and comparing time-spatial innovation dynamics through the lens of communities of practice. *Geoforum* (65): 338–350.

Murdoch, Jonathan (2006) *Post-Structuralist Geography: A Guide to Relational Space.* London, Thousand Oaks, CA and New Delhi: Sage.

Neff, Gina; Stark, David (2004) Permanently beta: Responsive organization in the internet era. In Philip N. Howard (Ed.): *Society Online: The Internet in Context.* Thousand Oaks, CA: Sage, pp. 171–188.

Nölke, Andreas (2009) Finanzkrise, Finanzialisierung und vergleichende Kapitalismusforschung. *Zeitschrift für Internationale Beziehungen* 16(1): 123–139.

Nolting, Michael; Mietzner, Dana (2010) Dienstleistungen in der roten Biotechnologie. In Dana Mietzner, Dieter Wagner (Eds.): *New Market Intelligence. Identifizieren und Evaluieren von Auslandsmärkten für Dienstleistungen in der roten Biotechnologie.* Wiesbaden: Gabler Verlag, pp. 5–44.

Nordhaus, William (2011) Designing a friendly space for technological change to slow global warming. *Energy Economics* 33(4): 665–673.

Oberhauser-Aslan, Heide (2014) SAP-Mitbegründer Dietmar Hopp: 'Für Biotech wird es keine zweite Milliarde geben'. *The Wall Street Journal*, 26 September. Available online at www.wsj.de/nachrichten/SB10553624399357934584840458017583169136548, 1 October 2015.

Obstfeld, David (2005) Social networks, the tertius iungens orientation, and involvement in innovation. *Administrative Science Quarterly* 50(1): 100–130.

OECD (Ed.) (2005) *A Framework for Biotechnology Statistics.* Paris.

Padgett, John F.; Ansell, Christopher K. (1993) Robust action and the rise of the Medici, 1400–1434. *American Journal of Sociology* 98(6): 1259–1319.

Paul, Stuart; Whittam, Geoff; Wyper, Janette (2007) Towards a model of the business angel investment process. *Venture Capital* 9(2): 107–125.

Pavitt, Keith (2005) Innovation processes. In Jan Fagerberg, David C. Mowery, Richard R. Nelson (Eds.): *The Oxford Handbook of Innovation*. New York: Oxford University Press, pp. 86–114.

Pike, Andy; Pollard, Jane (2010) Economic geographies of financialisation. *Economic Geography* 86(1): 29–51.

Pinch, Steven; Sunley, Peter (2009) Understanding the role of venture capitalists in knowledge dissemination in high-technology agglomerations: A case study of the University of Southampton spin-off cluster. *Venture Capital* 11(4): 311–333.

Pisano, Gary P. (2006) *Science Business: The Promise, the Reality and the Future of Biotech*. Boston, MA: Harvard Business School Press.

Popp Berman, Elizabeth (2008) The politics of patent law and its material effects: The changing relationship between universities and the marketplace. In Trevor Pinch, Richard Swedberg (Eds.): *Living in a Material World: Economic Sociology Meets Science and Technology Studies*. Cambridge, MA and London: MIT Press.

Powell, Walter W.; Sandholtz, Kurt (2012) Chance, Nécessité, et Naiveté: Ingredients to create a new organizational form. In John F. Padgett, Walter W. Powell (Eds.): *The Emergence of Organizations and Markets*. Princeton, NJ and Oxford: Princeton University Press, pp. 379–433.

Prevezer, Martha (2001) Ingredients in the early development of the US biotechnology industry. *Small Business Economics* 17(1–2): 17–29.

Prevezer, Martha (2008) Technology policies in generating biotechnology clusters: A comparison of China and the US. *European Planning Studies* 16(3): 359–374.

Reckwitz, Andreas (2012) Die Erfindung der Kreativität. Zum Prozess gesellschaftlicher Ästhetisierung. Berlin: Suhrkamp.

Rittel, Horst W.; Webber, Melvin (1973) Dilemmas in a general theory of planning. *Policy Sciences* 4(2): 155–169.

Rosiello, Alessandro; Parris, Stuart (2009) The patterns of venture capital investment in the UK bio-healthcare sector: The role of proximity, cumulative learning and specialisation. *Venture Capital* 11(3): 185–211.

Rosenkopf, Lori; Nerkar, Atul (2001) Beyond local search: Boundary-spanning, exploration, and impact in the optical disk industry. *Strategic Management Journal* 22(4): 287–306.

Rutten, Roel (2017) Beyond proximities: The socio-spatial dynamics of knowledge creation. *Progress in Human Geography* 41(2): 159–177.

Saxenian, AnnaLee; Sabel, Charles (2008) Roepke lecture in economic geography: Venture capital in the 'periphery': The new Argonauts, global search, and local institution building. *Economic Geography* 84(4): 379–394.

Schatzki, Theodore R. (2002) *The Site of the Social: A Philosophical Account of the Constitution of Social Life and Change*. Pennsylvania, PA: Pennsylvania State University Press.

Schudy, Simeon (2006) Jüngere Entwicklungen auf dem Risikokapitalmarkt für Biotechnologie in Deutschland. Kiel Institute for the World Economy (IfW) (Kieler Arbeitspapiere, 1270).

Schuldt, Nina; Bathelt, Harald (2009) Reflexive Zeit- und Raumkonstruktionen und die Rolle des Global Buzz auf Messeveranstaltungen. *Zeitschrift für Wirtschaftsgeographie* 53(4): 235–248.

Schumpeter, Joseph (1997[1911]) Theorie der wirtschaftlichen Entwicklung. Eine Untersuchung über Unternehmergewinn, Kapital, Kredit, Zins und den Konjunkturzyklus. 9th ed. Berlin: Duncker & Humblot.

Shiller, Robert J. (2015) *Irrational Exuberance*. Princeton, NJ and Oxford: Princeton University Press.

Smith, Adrian; Voß, Jan-Peter; Grin, John (2010) Innovation studies and sustainability transitions: The allure of the multi-level perspective and its challenges. *Research Policy* 39(4): 435–448.

Sokol, Martin (2013) Towards a 'newer' economic geography? Injecting finance and financialisation into economic geographies. *Cambridge Journal of Regions, Economy and Society* 6(3): 501–515.

Stark, David (2009) *The Sense of Dissonance: Accounts of Worth in Economic Life*. Princeton, NJ: Princeton University Press.

Stein, Axel (2014) The significance of distance in innovation biographies—the case of law firms. *Growth and Change* 45(3): 430–449.

Stockhammer, Engelbert (2009) The finance-dominated accumulation regime, income distribution and the present crisis. University of Economics and Business. Vienna (*Department of Economics Working Paper Series*, 127).

Strambach, Simone; Klement, Benjamin (2012) Cumulative and combinatorial micro-dynamics of knowledge: The role of space and place in knowledge integration. *European Planning Studies* 20(11): 1843–1866.

Stuck, Bart; Steingarten, Michael (2005) How venture capital thwarts innovation: The tech bubble saw an explosion of VC-funded start-ups—and a dearth of original ideas. In *IEEE Spectrum Online*, April. Available online at http://spectrum.ieee.org/computing/hardware/how-venture-capital-thwarts-innovation/0.

Sunder Rajan, Kaushik (2006) *Biocapital: The Constitution of Postgenomic Life*. Durham and London: Duke University Press.

Sunley, Peter; Klagge, Britta; Berndt, Christian; Martin, Ron (2005) Venture capital programmes in the UK and Germany: In what sense regional policies? *Regional Studies* 39(2): 255–273.

Szulanski, Gabriel (2003) *Sticky Knowledge: Barriers to Knowing in the Firm*. London: Sage.

Szyliowicz, Dana; Madsen, Tammy (2013) Waves of investing: Institutional dynamics in the venture capital sector. Paper to be presented at the 35th DRUID Celebration Conference, Barcelona, Spain, 17–19 June.

Thornton, Patricia H.; Ocasio, William; Lounsbury, Michael (2012) *The Institutional Logics Perspective: A New Approach to Culture, Structure, and Process*. Oxford: Oxford University Press.

Törnqvist, Gunnar (2004) Creativity in time and space. *Geografiska Annaler: Series B, Human Geography* 86(4): 227–243.

Trippl, Michaela; Tödtling, Franz (2011) Regionale Innovationssysteme und Wissenstransfer im Spannungsfeld unterschiedlicher Näheformen. In Oliver Ibert, Hans Joachim Kujath (Eds.): *Räume der Wissensarbeit. Zur Funktion von Nähe und Distanz in der Wissensökonomie*. Wiesbaden: VS Verlag für Sozialwissenschaften, pp. 155–170.

Truffer, Bernhard (2012) Environmental innovation and sustainability transitions in regional studies. *Regional Studies* 46(1): 1–21.

Ţurcan, Romeo V. (2008) Entrepreneur–venture capitalist relationships: Mitigating post-investment dyadic tensions. *Venture Capital* 10(3): 281–304.

Tushman, Michael L. (1977) Special boundary roles in the innovation process. *Administrative Science Quarterly* 22(4): 587–605.

Van de Ven, Andrew; Polley, Douglas; Garud, Raghu; Venkataraman, Sankaran (1999) *The Innovation Journey*. Oxford and New York: Oxford University Press.

Van der Zwan, Natascha (2014) Making sense of financialization. *Socioeconomic Review* 12(1): 99–129.

Van Osnabrugge, Mark (2000) A comparison of business angel and venture capitalist investment procedures: An agency theory-based analysis. *Venture Capital* 2(2): 91–109.

Vedres, Balázs; Stark, David (2010) Structural folds: Generative disruption in overlapping groups. *American Journal of Sociology* 115(4).

Vermeulen, Niki (2010) *Supersizing Science: On Building Large-Scale Research Projects in Biology.* Dissertation.com.

Vermeulen, Niki (2017) The choreography of a new research field: Aggregation, circulation and oscillation. *Environment and Planning A: Economy and Space* 50(8): 1764–1784.

Von Hippel, Eric (2005) *Democratizing Innovation.* Cambridge, MA: MIT Press.

Wallisch, Matthias (2009) Unternehmensfinanzierung durch Business Angels. Zur räumlichen Organisation des informellen Beteiligungskapitalmarktes. *Zeitschrift für Wirtschaftsgeographie* 53(1–2): 47–68.

Walsh, Vivien (2002) Creating markets for biotechnology. *International Journal of Sociology of Agriculture and Food* 10(2): 33–46.

Weber, Max (1980) Wirtschaft und Gesellschaft. 5th ed. Tübingen.

Weber, Barbara; Weber, Christiana (2007) Corporate venture capital as a means of radical innovation: Relational fit, social capital, and knowledge transfer. *Journal of Engineering and Technology Management* 24: 11–35.

Wenger, Etienne (1998) *Communities of Practice. Learning, Meaning and Identity.* Cambridge: Cambridge University Press.

Wenger, Etienne; McDermott, Richard; Snyder, William M. (2002) *Cultivating Communities of Practice: A Guide to Managing Knowledge.* Boston, MA: Harvard Business School Press.

Wójcik, Dariusz (2002) The Länder are the building blocks of the German capital market. *Regional Studies* 36(8): 877–895.

Wójcik, Dariusz (2009) Financial centre bias in primary equity markets. *Cambridge Journal of Regions, Economy and Society* 2: 193–209.

Wójcik, Dariusz (2011) *The Global Stock Market: Issuers, Investors, and Intermediaries in an Uneven World.* Oxford and New York: Oxford University Press.

Wong, Joseph (2011) *Betting on Biotech: Innovation and the Limits of Asia's Developmental State.* Ithaca, NY and London: Cornell University Press.

Wray, Felicity (2012) Rethinking the venture capital industry: relational geographies and impacts of venture capitalists in two UK regions. *Journal of Economic Geography* 12(1): 297–319.

Wu, Tim (2011) *The Master Switch: The Rise and Fall of Information Empires.* New York: Vintage Books.

Yakhlef, Ali (2010) The three facets of knowledge: A critique of the practice-based learning theory. *Research Policy* 39(1): 39–46.

Yeung, Henry Wai-chung (2005) Rethinking relational economic geography. *Transactions of the Institute of British Geographers* 30(1): 37–51.

Yin, Robert K. (2014) *Case Study Research: Design and Methods.* 5th ed. Los Angeles, CA: Sage.

Zacharakis, Andrew; Erikson, Truls; George, Bradley (2010) Conflict between the VC and entrepreneur: The entrepreneur's perspective. *Venture Capital* 12(2): 109–126.

Zademach, Hans-Martin (2009) Global finance and the development of regional clusters: Tracing paths in Munich's film and TV industry. *Journal of Economic Geography* (9): 697–722.

Zelizer, Viviana A. (1978) Human values and the market: The case of life insurance and death in 19th-century America. *The American Journal of Sociology* 84(3): 591–610.

Zeller, Christian (2002) Project teams as a means of restructuring research and development in the pharmaceutical industry. *Regional Studies* 36(3): 275–289.

Zeller, Christian (2003) Innovationssysteme in einem finanzdominierten Akkumulationsregime – Befunde und Thesen. *Geographische Zeitschrift* 91(3–4): 133–155.

Zeller, Christian (2008) From the gene to the globe: Extracting rents based on intellectual property monopolies. *Review of International Political Economy* 15(1): 86–115.

Zook, Matthew A. (2002) Grounded capital: Venture financing and the geography of the Internet industry, 1994–2000. *Journal of Economic Geography* 2(2): 151–177.

Zook, Matthew A. (2004) The knowledge brokers: Venture capitalists, tacit knowledge and regional development. *International Journal of Urban and Regional Research* 28(3): 621–641.

Index

Note: information in figures and tables is denoted by page numbers in *italics* and **bold**.